Welcome

to the

Microbiome

———————

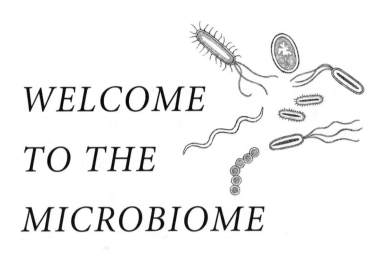

WELCOME TO THE MICROBIOME

*Getting to Know the Trillions of
Bacteria and Other Microbes
In, On, and Around You*

Rob DeSalle and Susan L. Perkins
Illustrated by Patricia J. Wynne

Yale UNIVERSITY PRESS/NEW HAVEN & LONDON

Yale University Press books may be purchased in quantity for educational, business, or promotional use. For information, please e-mail sales.press@yale.edu (U.S. office) or sales@yaleup.co.uk (U.K. office).

Set in Minion type by Newgen North America.
Printed in the United States of America.

ISBN: 978-0-300-20840-5 (hardback; alk. paper)

Library of Congress Control Number: 2015936985

A catalogue record for this book is available from the British Library.

This paper meets the requirements of ANSI/NISO z39.48 –1992 (Permanence of Paper).

10 9 8 7 6 5 4 3 2 1

Contents

Preface   vii

Acknowledgments   xv

CHAPTER 1. What Is Life?   1

CHAPTER 2. What Is a Microbiome?   36

CHAPTER 3. What Is On and Around Us?   69

CHAPTER 4. What Is Inside Us?   102

CHAPTER 5. What Are Our Defenses?   134

CHAPTER 6. What Is "Healthy"?   166

Epilogue   206

Glossary   209

References and Further Reading   217

Index   233

Preface

This book is about us—our human bodies, and especially the microbes that have taken up residence in and on every bit of them. Our species has been slow dancing with microbes for all of its existence. The common ancestor of all cellular life, human and microbial, appeared about 3.5 billion years ago as a single-celled microbe with no nuclear membrane to encase its genetic material. By one billion years ago, this ancestor had obtained a nuclear membrane, and its evolutionary trajectory had diverged from that of all the other single-celled organisms on the planet. During the next several millions of years, multicellular life flourished and diversified, and while many lineages went extinct, some survived, including our own. Living with our common ancestor during that time were a plethora of multicellular, and seemingly ever more complex, animals and plants. But one aspect didn't change: single-celled microbes continued their long and storied association with multicellular creatures, one of which would become *Homo sapiens.*

One hundred million years ago, our ancestor was a small mammal living side by side with dinosaurs, slews of insects, and a wide variety of plants, among other organisms. But this ancestor also lived with many species of microbes, and was most assuredly filled and covered from head to toe with them. Much later, ten million years ago, our ancestor was a much larger primate, also coevolving and coexisting with many microbes that lived in its guts, crevices, body cavities, and on its hairy surface. Still later, a million years ago, our ancestors were walking upright and were less hairy, but they were still populated with numerous microbes. There were as many as six species of humans walking the planet a million years ago, and each more than likely had its own unique cadre of coevolving microbes.

Although our ancestors 100,000 years ago could think about and place into context the world in which they lived, they knew nothing about microbes. They more than likely understood that there were features in their environment that occasionally influenced their bodies and made them sick. Sometimes what made them sick were poisons in plants, but more often than not it was microbe-spoiled food or microbe-infected dirty water, which they most likely learned to avoid. Ten thousand years ago, *Homo sapiens* populations started to change their lifestyles drastically, from hunting and gathering to more specialized strategies. This shift in lifestyles led to people living in villages and cities, and to many new interactions with microbes.

At this point in time, humans still had no concept of what caused illnesses such as diarrhea, fever, or infection after a wound. Even as recently as a thousand years ago, during the Dark Ages, illnesses caused by microbes were every bit as mysterious as they had been when we began our trek as a species. The Black Death, which was caused by a microbe, was treated entirely in a metaphysical context rather than scientifically. And there is ample evidence that other non-European cultures viewed their interactions with microbes in the same metaphysical or spiritual way.

Not until about one hundred years ago did we begin to learn about the previously unseen world of microbes. By this point, Antonie van Leeuwenhoek had invented the microscope, allowing microbes to be seen; vaccines were available for some diseases; Louis Pasteur had developed his sterilization approaches for microbes; and Robert Koch had come up with postulates for how microbes cause infections. The basis for a highly scientific approach to the microbial world had been established.

Starting a mere decade ago, at around the turn of the twenty-first century, this legacy of Pasteur and Koch had blossomed into a full-fledged scientific field. Antibiotics, antivirals, and the detailed clinical analysis of microbes were in place. DNA structure and the genetic code had been deciphered, and the first whole genomes of microbes had been sequenced. Researchers had begun to understand the broad diversity of microbial life using new technologies. The legacy of Pasteur and Koch was a century old and deeply entrenched in modern microbiology, and some illnesses were under control using the century-old paradigm.

Fast-forward to about one year ago, and we observe a shift in the paradigm of how we view the microbial world that stems from our starting to understand the breadth of microbial diversity and how this relates to ecology and our health. Factoring in the influence of our "microbiomes"—the particular assortment of microbes inside and on our bodies, as well as in the places where we spend our time, like our homes and our schools—is now commonplace. In fact, the Human Microbiome Project is well under way and instead of relying entirely on understanding microbially caused disorders as being controlled by a single microbe, we are beginning to understand the complexity of microbes' interactions with our bodies and the environments we inhabit.

This book is also about the recent paradigm shift in how we view the microbial world. To understand this shift, we first need to understand what life is and how it is organized on this planet. For the most part, the history of life follows a bifurcating pattern much like that first suggested by Charles Darwin. Although microbes violate this pervasive bifurcation by swapping genes back and forth in a process known as horizontal gene transfer, we can still pick up the signals of divergence and diversification of organisms over billions of years.

The great tree of life that Darwin envisioned in *On the Origin of Species* certainly serves as a wonderful backdrop for the last 3.5 billion years of evolution on Earth. But until about three decades ago, scientists thought, mistakenly, that there were only two basic kinds of cells on this planet—prokaryotes (organisms without a nuclear envelope) and eukaryotes (organisms with a nuclear envelope). It wasn't until the 1980s that researchers determined that there are actually three major kinds of cells—Archaea, Eukarya, and Bacteria. And although by the year 2000 only about seven thousand species of Bacteria and Archaea had been described and named, scientists in the 1990s and early 2000s proved that there are probably tens of millions of bacterial species on the planet, and maybe over a hundred million such species. Our understanding of the unique events of divergence through common ancestry has molded the techniques and strategies we use today in modern medicine and microbiology. That is, knowing that we have common ancestry with the very organisms that often make us sick has led to new approaches to how we interact with various environments and keep ourselves healthy.

The paradigm shift from a focus on single pathogenic organisms to understanding communities of organisms living in and on us has been possible because of a revolution in technology that allows us to "see" the extent of microbial diversity in small niches in and on our bodies. This technology, which we describe more fully in Chapter 2, uses microbial DNA sequences as signatures or "barcodes" for different species. As we will see, there are thousands of species of bacteria living on and in our bodies, most of them in commensal or mutualistic relationships with us. The microbes living on our bodies change with age, are different between the sexes, and are influenced by myriad environmental factors such as where we live and travel, the presence of dogs, or how active we are. Another important aspect of this technology is that it has illuminated the extent to which our bodies are swamped with bacterial genes—genes that are busy transcribing and translating their own proteins in tandem with our own DNA replication processes. To get a sense of the scale of the microbial action taking place, consider that our genomes hold a little over 20,000 genes, but the 10,000 bacterial species in and on our bodies have genomes that each average 2,000 to 4,000 genes. This means that we have about 20,000,000 bacterial genes coursing on and through our bodies—and that merely 0.1 percent or so of all of the different genes being transcribed and translated into proteins in your body right now are your own, a phenomenon also considered in detail in Chapter 2.

Chapters 3 and 4 describe the multitude of ecological niches our bodies have, and how microbes have taken up residence, shift their residence, and take advantage of their residence in these niches. These chapters should reinforce the idea that our bodies are crawling with microbes, yet that in general it's not the microbes that cause problems with our health, but rather disruptions in the natural ecology of our bodies that lead to illness. In other words, it is only when the coevolved ecological balance of our body's cells with the trillions of microbes living in and on us is thrown out of whack that pathogenicity arises.

Chapter 5 discusses what pathogenicity is and how microbes make us sick. Because we are now aware of the thousands of microbial species that live with us, a new way of interpreting pathogenicity is needed. Instead of single microbes causing infectious disease, we now have to consider the possibility that different species act together in this way. We also have to keep in mind that our bodies have coevolved to exist with microbes, and so have probably developed some defenses against them that have the potential to become pathogenic. That is, this new way of looking at the ecological balance of bacteria and our cells is also tempered by how our cells and bodies stave off infections. This chapter, then, also considers the immune system and how the ability of our cells and our bodies to resist infection has changed over the millions of years we have coevolved with microbes.

In addition to learning about the diversity of microbial life living in and on us, it is equally important to understand the interactions we have with microbes on an ecological or global scale. In fact, many of the therapies that we have created to combat pathogenic microbes are based on a very limited understanding of the microbial ecology of our bodies. Modern science has taught us some hard lessons about acting on our environment with this sort of limited information. For example, one of the prevailing strategies in biology for controlling pests is to introduce something to disrupt the pest. These introductions can be chemical, as with herbicides, or organismal, as with the introduction of the mongoose to control rat populations in certain areas of the world. These kinds of efforts have almost always resulted in unintended consequences such as herbicide-resistant insect populations. If we are to have any chance of controlling pathogens without causing additional harm, we will need to find new ways to examine and understand the overarching ecology of their habitats: our bodies. Complicating these questions is the fact that all humans do not have the same microbial makeup. Instead, a diverse mix of microbial communities interacts with us, which means that any clear understanding of our bodies will require us to consider population and cultural factors and their effects on our wider ecological communities.

Chapter 6 delves into what it means to be healthy and sick. Our definitions of sickness are better understood when we have a clearer picture of how microbes interact with us. Many of the interactions we have with microbes don't make us sick, and even more of the interactions that do make us sick can only be controlled by making us sicker. Current data on a lot of microbially induced pathogenesis indicate that eliminating a pathogen can also induce secondary illnesses and other pathogenic responses that we often cannot predict. Our contention in Chapter 6 is that by using appropriate model systems and conducting well-thought-out experiments we can make some headway toward creating therapies for illnesses caused by pathogenic microbes. We hope that readers will come away from this discussion with a better sense of their bodies and how microbes interact with them, an appreciation for the importance of these unseen inhabitants to human health and sense of well-being, and tools for making informed decisions about the health of their microbiomes and themselves.

So, sit back and let us introduce you to the trillions of organisms living in, on, and all around your body.

Acknowledgments

We would like to acknowledge the American Museum of Natural History Exhibitions staff, in particular David Harvey, for their brilliant development of a major exhibition on microbiomes. We also acknowledge the creative writing staff of the department headed by Lauri Halderman, and specifically Martin Schwabacher, for their help in bringing the initial ideas for this exhibition to fruition.

We wish to thank our colleagues Vivian Schwartz and George Amato, who read early drafts of the manuscript, and Stephen Gaughran for his suggestions for the glossary. We appreciate as well the expert editing and input of Jean Thomson Black, our editor at Yale University Press, and the masterful technical assistance of Samantha Ostrowski.

A great and heartfelt note of appreciation also goes to Patricia Wynne, who prepared all the line illustrations that enrich and enliven our writing.

Finally, Rob DeSalle acknowledges the support of his wife, Erin DeSalle, without whose encouragement this book would not have been possible.

CHAPTER 1

What Is Life?

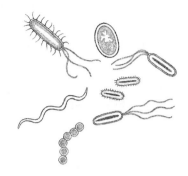

You are not what you think you are. Instead of the form staring back at you when you look in the mirror, what you should imagine is your body as a collection of multiple dynamic ecosystems made up of very tiny, and very biologically diverse, organisms. Most of these organisms exist in a landscape of varied and dynamically changing environments. For instance, your armpit might be a dry and rather cool ecosystem after sitting quietly in an air-conditioned room, but after you exercise, this habitat will be hot and steamy. People tend to feel adversarial toward the many organisms that inhabit and live on our bodies, but the more we learn, the more we will realize that this attitude can be misguided and even dangerous for our health.

Let's start by making distinctions among the many common names given to microbes—the organisms that are so small that we must use microscopes of one sort or another to see them. The most common and probably most familiar name for microbes is "germ," but this seems inaccurate because only a small number of microbes cause disease. "Bug" is also not a good moniker because it confuses microbes with a common (and sloppily applied) slang word for insects. Some people might think of microbes as "bacteria," and they would be right in part, because "bacteria," when not capitalized, does refer to microbes nonspecifically. But when the word bacteria is capitalized (Bacteria), it means a huge group—known to biologists as a domain—of single-celled organisms with a set of shared attributes and a common ancestor, one different than that of all other life on our planet.

Why be so specific? Because there are two other domains of organisms that don't share this Bacterial common ancestor and hence aren't called Bacteria: the domain of single-celled microbes known as Archaea, and the domain of single-celled or multicellular organisms called Eukarya. These, then, are the three major groups of living organisms on this planet discovered so far. Throughout this book we will use the word microbe when we are talking generally about any single-celled life that is a member of the Bacteria, Archaea, or Eukarya domains. When we use the term Bacteria or Archaea, we are referring to two specific taxonomic groups of single-celled organisms without nuclei, those membrane-bound structures that contain the chromosomes of organisms. Perhaps the most recognizable member of the Bacteria group to readers will be a species called *Escherichia coli* or *E. coli*, the species that resides comfortably, for the most part, in our guts. Archaea are a bit more obscure, but there have been examples of some members of this domain in the news, such as *Haloquadratum walsbyi*, a bizarre rectangular-shaped organism.

Taking a Closer Look

Given the importance of our microbiome, understanding human health could be considered a problem for a biodiversity expert, whose first step would be to account for what is in our microbial ecosystem. The biologist E. O. Wilson once said, "When you have seen one ant, one bird, one tree, you have *not* seen them all." We need a precise assay of what lives in us, on us, and in our crevices in order to understand the ecosystem that is the human body.

Where should we start this investigation? At the smallest, most detailed level: the subatomic scale. Carl Sagan once called us, and everything around us, "star stuff," and Neil deGrasse Tyson has pronounced that "we are not figuratively, but literally, stardust." Everything on this planet is made of stardust, and stardust is made up of particles, some of them extremely small—molecules and atoms—as well as a few still theoretical, such as tachyons. It's how the particles are arranged, and how they stick together, that makes us what we are, what makes a microbe what it is, and a lamppost what it is.

How did these particles come to be here on Earth? Most particle physicists would point out that there is a cosmic connection among all animate and inanimate objects on this planet, and indeed, this universe. All matter originated with the Big Bang, an event that occurred about 13.5 billion years ago. How we know this and what it has produced is a subject for an entire cosmology course. Suffice it to say that the event produced the small particles, and in turn, the elements, that are so important to our understanding of life on this planet. And for those scientists who consider themselves astrobiologists, there is a strong tendency to think that other solar systems and galaxies have the same elemental makeup.

Yogi Berra, the famous baseball player and homespun philosopher, once said, "When you come to a fork in the road, take it." This seemingly funny advice reveals how we look at phenomena in the universe. Humans like to name things. In so doing, we think in terms of splitting or bifurcating. If one object is called "Onk," then anything that isn't Onk has to be called something else, and so on. It is a very logical way to think and to make sense of the world. We clearly like to name objects, because names (nouns) are a part of every language on this planet. Moreover, we are not the only organisms on this planet that engage in naming. For example, African vervet monkeys have specific cries they make depending on whether a bird of prey, a leopard, or a snake is near. Prairie dogs have a very specific, noun-based way of communicating. And even Bacteria communicate with each other in a "nouny" way, using a process called quorum sensing.

A paleontologist colleague of ours who is something of an Indiana Jones wannabe once pointed out that bifurcation is built into the basis of every way that organisms communicate with the outside world. The first and most basic bifurcation is, "Will the organism kill me or not?" The next level of recognition or naming is "If it can't kill me, then can I eat it?" And the third level of communication is "If I can't eat it, can I have sex with it?" All other naming decisions are made after these three questions have been answered.

Humans have developed a very complex way of naming. This is the beauty of the languages that continue to evolve with our cultures. For the task of classifying the various organisms on Earth, for example, we have created multiple names corresponding to the different levels of complexity that can be considered. So for instance, even though all objects in the universe are made of elemental particles, the names of the elemental particles have little to do with the names of the atoms that these elemental particles make up. Similarly, the names of the 115 different kinds of elements have little to do with the names of objects like rocks or organisms. And thanks to a Swedish scientist named Carl von Linné (also known by his Latinized name Carolus Linnaeus), the names of living organisms have been classified in neat bifurcating groups that mirror the bifurcation that is so important in nature. Linnaeus's binary way of classifying things has led to a plethora of mnemonics for remembering the levels of classification that arose from his thinking, such as "Dear King Philip come over for good soup" (for "domain, kingdom, phylum, class, order, family, genus, species").

For instance, we are part of a genus, *Homo,* and a species, *sapiens* (Latin for "wise"). The genus *Homo* also includes many different species that have gone extinct, like *Homo neanderthalensis, Homo erectus,* and *Homo ergaster.* All genera that are closely related to our genus, but are not ours, are part of a family called Hominidae or great apes; these include *Gorilla, Pan, Pongo,* and several extinct species. The naming process goes on for several other levels, until an organism has several different names that tell other humans (especially those biologically inclined) what they are. So our full name is Eukarya, Animalia, Bilateria, Deuterostomia, Amniota, Chordata, Mammalia, Primates, Hominidae, *Homo sapiens.* Each name is applied according to strict conventions and rules, and tells the listener or reader something about the origin of the organism.

For the bacteria we consider in this book, the naming protocols are quite different from those used for us humans. For example, the name of the gut bacterium on which we depend for proper digestion of food, *E. coli,* has the full name Bacteria, Proteobacteria, Gammaproteobacteria, Enterobacteriales, Enterobacteriaceae, *Escherichia coli.* Note that the name is somewhat simpler and has fewer levels than the name for humans, but this is partly because the bifurcation for bacteria is less well understood than for mammals or eukaryotes. We could very easily have as many names for a species of bacterium as we have for us, if we recognized the distinguishing properties of the various bacteria more precisely and could use those properties to classify the bacteria in a bifurcating way. In fact, part of the reason that we have fewer names for single bacterial species is that we have named only seven thousand species, whereas with Eukaryotes we have names for over 1.7 million species. Why is this disparity important? It has a lot to do with how scientists in the future will confront the ways that microbes interact with the ecosystem of the human body.

Substitution Codes

The objects we see around us are filled mostly with space, because they consist of atoms, which are made of small particles clumped together in what is called a nucleus. The nucleus is encircled by even smaller particles orbiting at huge distances around it called electrons. For the most part, there is nothing in the space between the orbiting electron and the nucleus. Two atoms can stick together through a number of mechanisms, the most common of which is to share orbiting particles. If atoms could not stick to one another, then we would have a simple universe where the various kinds of atoms produced by different numbers of particles would be the universal citizens, instead of the myriad forms of animate and inanimate things we do see around us.

Some atoms stick together in ways that promote lattices or clumps. For instance, sodium atoms will stick to chlorine atoms in a one-to-one ratio to make a crystal lattice called sodium chloride, which is more commonly known as table salt. By contrast, atoms of carbon, phosphate, hydrogen, nitrogen, and oxygen can stick together in more linear ways. If these five types of atoms configure in a certain fashion, they can form what are called nucleotides, which are common to life on our planet. There are five kinds of these nucleotides, depending on the structure of the nitrogenous base they contain—guanine (G), adenine (A), thymine (T), cytosine (C), and uracil (U). Deoxyribonucleic acid (DNA) consists of linear arrangements of nucleotides with deoxyribose sugar, phosphate, and the four bases guanine (G), adenine (A), thymine (T), and cytosine (C). Ribonucleic acid (RNA) is a linear arrangement of nucleotides with ribose, phosphate, and the four bases guanine, adenine, thymine, and uracil (G, A, T, and U). If oxygen, hydrogen, carbon, nitrogen, and sulfur combine in certain ways different from those used to form nucleotides, they can instead form amino acids. There are twenty different configurations of these atoms for forming amino acids (actually there are a few more, but the twenty are largely used by our cells and microbial cells). The alphabet for the twenty amino acids is shown in the following table:

Glycine = G	Histidine = H
Alanine = A	Arginine = R
Valine = V	Asparagine = N
Leucine = L	Glutamine = Q
Isoleucine = I	Glutamic acid = E
Serine = S	Aspartic acid = D
Threonine = T	Phenylalanine = F
Cysteine = C	Tryptophan = W
Methionine = M	Tyrosine = Y
Proline = P	Lysine = K

Nature has evolved such that RNA has become an intermediary between the genetic material (DNA) and the workhorses of our cells made of amino acids called proteins. Since DNA and RNA have the same basic elements (they both have G's, A's, and C's, and DNA has T's whereas RNA includes U's), DNA is said to be *transcribed*, or rewritten, into RNA. The amino acid language is different from the nucleotide language, so RNA is said to be *translated* into amino acids that then make up proteins.

To make a protein, then, we have to go from storing the genetic information in the four letters of DNA to the four letters of RNA.

This transcribing step is easy to understand, because it is a simple one-to-one exchange and follows the base pairing rules of nucleic acids. G pairs with C, and T (or U) pairs with A. The only hard thing to remember is that the transcribed RNA is said to be "antiparallel and complementary" to the DNA of the gene. But this problem is just a matter of keeping track of orientation of the DNA and the RNA. Once the DNA has been rewritten as RNA, the RNA needs to be translated into protein. Here is where a trick of nature called the genetic code comes in.

Codes and cryptography have been an important part of keeping secrets for hundreds of years. One of the easiest of all codes to break is the simple substitution code. This is the kind of code that is created by simply substituting one letter for another. It's the kind of code that Ralphie, in the classic American movie *A Christmas Story*, uses to decipher, with a decoder ring, an apparently important message he hears over the radio. (To his disappointment, the message he decodes is "Be sure to drink your Ovaltine.")

Although protein synthesis uses a sort of a substitution code to translate from RNA to protein, we can't have a simple one-to-one exchange, because there are twenty amino acids and only four nucleic acids. We could use two nucleic acids to represent a single amino acid, but we would still be short some coded doublets, because four nucleic acids can only make sixteen unique pairs (GA, GT, GC, GG, AT, AG, AC, AA, CT, CA, CG, CC, TA, TT, TC, and TG). If we use three bases to code for a single amino acid, then we would have sixty-four possible codes and only twenty amino acids, but then it seems as though we would have too many code triplets or codons. In the 1950s and 1960s, however, researchers determined that nature has taken up the triplet code system. The surprise is that some amino acids are coded for by more than one triplet nucleic acid "word." For instance, the four codons CCC, CCT, CCG, and

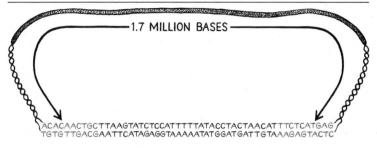

Figure 1.1. Diagram showing the sequence of a miniprotein gene called *blpC*, from the *Bacillus subtilis* genome. The helical ended line represents the entire *B. subtilis* genome of approximately 1.7 million base pairs. The *blpC* gene itself is thirty-three base pairs long and codes for a miniprotein eleven amino acids in length.

CCA all code for the amino acid proline (P). The use of multiple codons for a single amino acid is called the redundancy of the genetic code.

Let's take what we have learned and look at one of the smallest genes around in a bacterial genome. It is a gene for a miniprotein called blpC and is from *Bacillus subtilis,* a species of bacterium we will revisit later. The genome sequence around the small gene is shown in Figure 1.1. There are 1.7 million or so bases in the circular *B. subtilis* chromosome, so while we are showing about forty bases, imagine the enormity of the rest of the chromosome surrounding the *blpC* gene.

Transcribing DNA to RNA requires us to think in reverse and at the same time complement the A's with T's and G's with C's.

By reversing the sequence we go from

TTAAGTATCTCCATTTTTATACCTACTAACAT

to

TACAATCATCCATATTTTTACCTCTATGAATT

and by taking its complement while substituting G's for C's and U's for A's and C's for G's and A's for T's we get the following sequence of the RNA that codes for the protein:

AUGUUAGUAGGUAUAAAAAUGGAGAUACUUAA.

If we use the single-letter amino acid alphabet listed earlier as our guide, we discover that the amino acids that this RNA encodes are M R N V S R Y K N G D T – , with the "M" resulting from the base triplet "AUG," the "R" from the base triplet "AGA," and so on.

Before we go on, we should note a few important things about translation from RNA into protein. First, this protein sequence starts with an M. The M stands for the amino acid methionine. It turns out that nearly every protein starts with an M. If we liken a protein sequence to a sentence, then the presence of methionine (M) indicates the capitalization of the sentence or the start of a protein. Not all methionines are at the beginning of proteins, so we need to be careful, but the presence of an M is usually the first thing that scientists look for when they look for genes in the mess of a genome sequence. Methionine (M) is a rather unique amino acid with respect to the genetic code in that there is only one three-nucleotide codon that represents it in proteins—AUG. This means that the genetic code is *not* redundant for methionine. Next, note that there is a dash at the end of the protein. This indicates that the protein has stopped, and indeed the three-letter DNA code for this dash (there are three of them in general) is called a "stop codon." In the case of our blpC protein, the stop codon is UAA and we separate this codon from the rest of the RNA sequence.

In our sentence analogy, the three-codon stop codons would represent periods at the end of the sentence. When researchers look for genes in genome sequences, they usually start with the capitalization codon M and try to find stop codons further downstream in reasonable places.

Now note that there are thirty-nine nucleic acids (GATC), twelve amino acids, and one stop codon making thirteen codons, reflecting the triplet nature of the genetic code. The second codon is AGA, and this codon stands for arginine, whose amino acid abbreviation is R. This codon is separated by a space from the rest of the RNA sequence.

DNA molecules can be made of chains of nucleotides millions of nucleotides long. These long chains are called chromosomes. So for instance, most *Bacillus subtilis* have a single circular chromosome. The strain of *B. subtilis* on which we have been focused has about 1.7 million G's, A's, C's, and T's in its single circular chromosome, and this makes up the genome of this bacterial strain. There are 1,914 genes in the genome of this strain of *B. subtilis*, of which *blpC* is one. While *blpC* is only 69 nucleotides long, the 1,913 other genes in the genome of this strain are, on average, about 700 nucleotides long.

Thus far we have not discussed the intricate system of proteins, enzymes, and molecular structures that implement the transcription of DNA to RNA and the translation of RNA to protein. This system is also integral to learning basic biochemistry, because it is the protein makeup of cells that makes them what they are.

Why Are Ancient Bacteria Important?

Determining how life on this planet arose is difficult. It is so difficult that some scientists have simply given up and suggested that life may have arrived on this planet as a result of transport by means of celestial bodies such as meteors or on comets that collided with the Earth. This idea is known as the panspermia hypothesis. Other researchers take a grittier approach and assume that nucleic acids, amino acids, and other complex molecules as well as viruses and cells arose on the planet through evolutionary processes. Some progress has been made on this front, in part because many researchers think that RNA is an important intermediate in how life arose on the planet. They call the period of time when RNA was king "the RNA world." Their inferences about RNA as the first step in complexity are based on what we know about RNA in biochemistry. RNA seems to be able to perform by itself a lot of the processes that life needs such as catalyzing reactions and replicating itself. If we accept the hypothesis of an RNA world, then about 3.5 billion years ago the RNA world flipped to a DNA, RNA, and protein world. If we accept the panspermia hypothesis, then cells more than likely arrived on the planet 3.5 billion years ago fully equipped to evolve. Either way, it is clear from fossil evidence that life on this planet arose about 3.5 billion years ago, only about 1 billion years after the planet Earth itself formed.

There are a few species of microbes that are large enough to see with the naked eye, but the grand majority can only be observed through microscopes. One can take a bacterial culture and smear it on a glass slide and observe it under a microscope. If the bacteria are kept alive, then depending on which bacteria were used, they will appear under the microscope to be "swimming" around, darting around, or just slugging about. Another way to view microbes is kill them, stop the processes in the cell, and embed them in a tiny resin block. Researchers can then slice the bacteria embedded in the plastic into very thin slices. Resin is used because it immobilizes the bacteria and sticks to them really well. It is also translucent, so the slices can be placed on a slide and observed through a microscope with the aid of a light from underneath. This approach is how the very fine structures of living microbes are determined.

Fossil bacteria are embedded in rocks, somewhat like bacteria embedded in resin. But rock is much harder than resin, so recovering thin slices of bacteria embedded in rocks is very difficult. In the 1980s, however, researchers in California, led by J. William Schopf, were able to take a kind of rock called chert, which is relatively soft, and cut it into thin slices that they could observe under a microscope. The chert was determined to be about 3.5 billion years old, and to contain the remnants of bacterial cells. Since then, hundreds of thinly sliced rocks have been examined, and a whole menagerie of bacterial life has been found at regular intervals in the fossil record. What this means is that the 3.5 billion–year-old bacteria that were first discovered in the 1980s were not an anomaly, but rather examples of bacteria that have been dispersed all over the Earth for the past 3.5 billion years.

Thanks to the pioneering work of the Illinois-based biologist Carl Woese, we now know that there are three major kinds of cellular life on the planet. We have already written about humans as Eukaryotes and about some single-celled organisms called Bacteria. But there is also a category of single-celled organisms that are different from Bacteria and Eukaryotes called Archaea, or more formally, Archaebacteria. These single-celled organisms are so named because they appear to live in what researchers think might have been primordial environments, such as places that are under extreme pressure, or landscapes that are very hot and perhaps have high concentrations of chemicals such as sulfur or methane. In fact, because many Archaea live in these extreme environments—for example, thermal vents on the ocean floor, or hot springs in Yellowstone Park—they have been labeled as extremophiles. But there are Archaea that live in not-so-extreme places such as our mouths, which despite seeming like an extremely bad place to live turns out to be an appealing environment to many microbes. In addition, there are single-celled organisms that aren't Archaea that live in extreme environments like hot springs. The most famous of these is a bacterium called *Thermus aquaticus*. This species of bacterium makes an enzyme that has become an important part of a technique used by scientists called polymerase chain reaction, or PCR.

If that weren't enough, we have yet to discuss viruses. Viruses are truly an oddity. Amongst the smallest viruses are flu viruses with eight genes, or papillomaviruses (viruses that cause cervical cancer), which also have eight or so genes. There are some rather large viruses such as *Mimivirus*, with almost a thousand genes. But the largest virus found so far is called *Pandoravirus*, which has more than 2,200 genes. These two viruses actually have more genes than some Bacteria and Archaea. Viruses act to disrupt the normal processes of the cells in which they exist (called host cells) and co-opt the usual replication machinery of the cell to make copies of themselves. A typical virus will have an outer protein shell, called a capsid, which protects its genetic material. On the surface of the capsid are tiny molecules that the virus uses to attach to cells. Then the virus does the one thing that all viruses do very well, which is to override the host cell's own genetic machinery so that it will replicate the virus (Figure 1.2).

Many scientists don't feel that viruses should be categorized as life. For this reason, and because viruses more than likely do not share a common ancestry with other organisms, they are not usually included in the tree of life. Interestingly, viruses themselves probably do not all come from the same ancestor. There are two lines of evidence for this idea. First, researchers know the names, numbers, and function of genes in most of the viruses with which we are familiar, and there is no single gene that is found in all viruses. Closely related viruses have similar genes, but viruses that have very different functions and habitats have very, very few genes in common. By contrast, at least 7 percent of the genes in a typical bacterium are shared with eukaryotes, indicating a deep and traceable common ancestry.

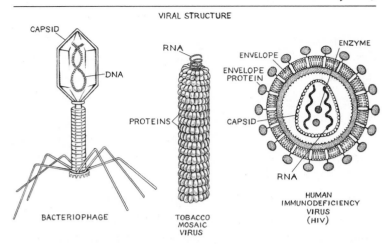

Figure 1.2. Schematic drawings of typical phages or viruses. The drawing on the left shows the head and tail structure of a bacteriophage, a kind of phage that infects bacteria. The figure in the middle shows the structure of the tobacco mosaic virus, which infects tobacco plants. The figure on the right shows the structure of HIV.

The second line of evidence arises from the broad differences in the kinds of viruses in existence and how they work. For cells, there is a single kind of nucleic acid that carries the genetic material—double-stranded DNA. But viruses have several methods for packaging their genetic material. Some viruses indeed employ double-stranded DNA, but others work with a single strand of DNA. If that weren't odd enough, there are also viruses that use RNA as the carrier of genetic material. Why not? RNA follows the same rules of pairing as DNA does, and so can carry and pass along the necessary genetic information. In fact, just as there are single-stranded and double-stranded DNA viruses, there are also single-stranded and double-stranded RNA viruses. Within the double-stranded RNA viruses there are also two categories of viruses—those that use one strand of the double-stranded RNA as a direct molecule for translating from RNA to protein, and another that requires that RNA be synthesized as an intermediary step before the proteins are generated. These disparate ways of packaging, transferring, and utilizing the genetic material suggest a major disconnect when it comes to viral origins.

No one has yet found a fossil virus in rocks, though some researchers continue to look for traces of viruses in rocks that also contain fossil microbes. One possible way to find a fossil virus or microbe might be, believe it or not, to find it intact. A recent dig in 2014 from Siberian ice released a virus that is 35,000 years old, and in 1995 Raúl Cano, a California biologist, claimed to have isolated microbes from amber fragments that are millions of years old. If viruses have been parasitic with bacteria for long periods of evolution, any microbe isolated from old amber might also contain viruses. To our knowledge, however, no one has tried to extract a virus from those microbes.

Genome scientists called viral paleontologists use the fact that viruses often embed their genes into the genomes of their hosts to tease out the evolutionary history of viruses. When viruses embed their genes in this way, the genes go into an inactive "hibernation," but are still replicated as part of the host chromosome. Instead of being embedded in rock or in amber, then, these viral genes are embedded in eukaryotic genomes. Viral paleontologists compare the embedded viral DNA with DNA from living viruses to understand how they have changed over time. They use a method called a molecular clock to estimate how old the embedded fossil viral genes are. A molecular clock takes advantage of the idea that DNA mutates at a regular rate, kind of like the ticking of a clock or the radioactive decay of isotopes. By counting the number of changes among viruses, the molecular clock can tell researchers how long a virus has been embedded in the host genome. Given some of the tight associations of viruses with cells, and the fact that some viruses are RNA-based, it is more than likely that viruses are quite old too, perhaps even as old as cells.

This discussion of viruses brings us to the general question asked in the title of this chapter, "What is life?" As noted previously, many researchers refuse to consider viruses living organisms. So how do we define life to exclude viruses? Most researchers use a simple thought problem to implement the definition, whether they know it or not. Imagine a cell left alone in some nutrients. There are no other cells around it. Chances are that cell will start biochemical processes within its walls and will eventually replicate itself, because it has the genes and biochemical machinery to do so. Now imagine a virus in the same situation—by itself in a pool of nutrients. Nothing happens. Not even a single biochemical reaction within the wall or capsid of the virus will occur. It cannot ever replicate itself if it remains alone. The cell is alive because it can carry on biochemical processes by itself and can replicate. The virus is not alive because it can do neither. This is a pretty persuasive and objective definition of life, and one that we will use throughout this book. But it is important to emphasize that viruses are on the edge of the defining line of life. By slightly changing our thought experiment or our definition of what is and is not life, viruses can very easily be considered to be alive.

Bifurcation or Trifurcation?

There are many ways that the three forms of cellular life could have arisen. The most obvious mode would involve an early splitting of the common ancestor of all cells into two daughter lineages, and then a second later split. This process is how Darwin postulated that three species would diverge from one another. But there might instead have been many splits in the primordial soup, with all but three lineages undergoing extinction, to leave us with the three major lineages of cellular life that we now observe. Still another possible explanation brings to mind the Arthur C. Clark science fiction classic *Rendezvous with Rama*. In this novel everything occurs in threes. That is, we might need to consider the possibility that the early branching of the three major cellular lineages was a trifurcation, looking like a pitchfork with three prongs.

To understand how scientists have recovered the important historical events that led to the three major groups on this planet, we need to understand something about the characteristics and genomes of organisms and how they have been named in the past. Take, for example, the name Prokaryote, which may be familiar to many readers. Notice that it is capitalized and hence means that it supposedly refers to a group of organisms that have a single common ancestor to the exclusion of others. Prokaryote is a hypothetical group made up of single-celled organisms that lack a membrane around their DNA. In the real world of organisms on our planet this would include all Bacteria and all Archaea together. If this hypothesis is correct, then the third major kind of cell, comprised of those with a membrane around their DNA and called eukaryotes, is the odd man out. This is the hypothesis that most of us were taught in high school and college and it is one that we can test by understanding the relationships of Bacteria, Archaea, and Eukarya.

To figure out these relationships, we first need to solve a simple mapping problem. In a set of three species, if you suspect that two of them are more closely related than either is to the third, you need an outside frame of reference to prove it. In other words, you need to be able to root the three species. This is because without a root, the relationships of the three species you are worried about will always look like the tree in Figure 1.3.

If you root the tree between Eukarya and the X, this would mean that Bacteria and Archaea are more closely related to each other. If

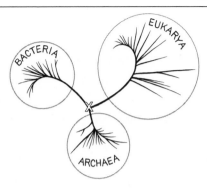

Figure 1.3. Phylogenetic network showing the relationships of the three great domains of life—Archaea, Eukarya, and Bacteria. The X indicates the junction of the three domains in the network. If the tree is "rooted" on the X, then all three domains are equally related to each other.

you root the tree between Archaea and the X, then Bacteria and Eukarya would be each other's closest relatives. And if you root the tree between Bacteria and the X, Archaea and Eukarya would be each other's closest relatives. But what is the X? That is, what do we use to root this tree? Because we have said these are the only three kinds of living cells on the planet, we have nothing on this planet to use as a root. Another classification problem arises when we consider the sex lives of microbes and the way they exchange DNA. To solve this problem, we need to know a little more about the genomes of life on the planet.

Microbial Genomes: Sharing and Duplicity

Another problem complicating the task of classifying these living cells arises when we consider microbes' sex lives—or rather, gene-swapping activities. Poor Bacteria and Archaea; they really don't have sex. In general, they propagate by replicating their DNA and splitting into two new daughter cells. Forget about Dolly the sheep and the hundreds or so other animals that have been cloned in the last decade: Bacteria are the ultimate clones. If they made perfect copies every time, however, they would be horrible at adapting to new environmental challenges, because in order to adapt, variation is needed—and by far the most efficient way to introduce variation is to absorb some DNA from elsewhere and hope that it might help the cell when faced with a novel environment. There is a gentle balance required here, because if a cell keeps sucking up foreign DNA, there will eventually be a point where the absorbed DNA will become a detriment to the fitness of the microbe.

There is a great deal of controversy surrounding whether or not a "tree of life" exists for Bacteria and Archaea. This controversy arises because both of these major kinds of cells are rather promiscuous when it comes to exchanging DNA. Since Bacteria don't have sex the way eukaryotic cells do, they have evolved several mechanisms for acquiring foreign DNA, such as conjugation, transformation, and transduction. These modes of moving DNA around different species of Bacteria and Archaea are called horizontal gene transfers, or HGTs. This promiscuity means that, when HGT from one organism to another occurs, the overall relatedness of organisms as indicated by that piece of DNA is disrupted. The role of HGT in understanding the relatedness of microbes has prompted some researchers to herald the demise of Darwin's strictly bifurcating tree of life and to claim that a comb of life is a more appropriate way to look at this process. We disagree with these claims. Scientists do have a pretty good idea of how Bacteria evolved and how they are related to each other. Giving up and deciding that we have a comb leaves us with no frame of reference to organize the events that have occurred in the past 3.5 billion years of life on the planet, even the HGT events that some researchers think are detrimental to reconstructing the tree.

The Japanese scientist Susumo Ohno's book *Evolution by Gene Duplication,* written in the early 1970s, describes these processes in detail. To follow Ohno's argument, recall that the genomes of organisms are made up of long strings of DNA compartmentalized into smaller clumps called chromosomes and then into even smaller snippets called genes. The genes have different arrangements of guanine, adenine, thymine, and cytosine, and these differences are what code for the different proteins in cells. Microbes have evolved hundreds of thousands of kinds of genes that make proteins, which have literally hundreds of thousands of functions. But there is a set of genes, the "core" gene set, that is common to all microbes and is minimally needed to make a bacterium.

One way to figure out the core set of genes of microbes is to look at species that have lost most of their genes. Why would an organism lose genes? Because they sometimes become parasitic. The Bacterial genera *Buchnera* and *Wolbachia,* for example, have evolved to invade eukaryotic cells and live in the cytoplasm of those cells. The eukaryotic cell takes over many of the genetic functions of the bacteria's genomes, such as the synthesis of components for their cell membranes, and so these bacteria lose genes left and right. The record for the smallest number of genes in one of these parasitic Bacteria is 182, and some other parasitic or facultative species are pretty close to that size. The Archaea examined so far all have genomes over 1,400 genes, so they don't seem to trim themselves down the way that facultative Bacteria do. In fact, the largest archaeal genome is *Methanosarcina,* which thrives on methane and has 4,540 genes, or about the average size of a bacterial genome. By contrast, the largest bacterial genome is found in a soil microbe, *Ktedonobacter:* its genome has 11,400 genes, which is only two thousand genes smaller than the genome of a fruit fly.

Another way to figure out the core set of genes involves pinning down which genes overlap in all bacteria. Using experiments designed to do just that, researchers have determined that there are about 180 genes in the core set. Offering further proof that these are truly the core set, scientists have artificially synthesized these 180 genes into an artificial chromosome, placed them into an otherwise empty bacterial cell, and, lo and behold, discovered that the synthetic bacteria can function and reproduce on their own. The ultimate goal of researchers who work in this new area of biology (called synthetic biology) is to produce an artificial membrane and do the same trick without the empty bacterial cell.

Ohno's idea was that once a particular set of successful genes evolved, the combined function of those genes would be well suited to the lifestyle of the organism. If the organism were to be challenged by a new environment that required some novel gene function, however, then the organism would need to adapt genetically to that environment. Ohno suggested that the simplest way an organism could acquire the new genes it needed to succeed in the new environment would be to duplicate the existing genes and then let those copies adapt.

Ohno didn't come up with this important hypothesis in a vacuum. He worked mostly on fish, a group of organisms in which some species have larger and larger numbers of chromosomes. In fact he and other researchers, who were working on plants, recognized that often closely related species have exact multiples of chromosomes. For instance, if one species has two chromosomes, its related species might have four chromosomes, eight chromosomes, or even sixteen chromosomes. The species with the higher multiples of chromosomes are called polyploids. What all of this means is that some genes in the genomes of organisms will have patterns that more closely resemble gene clusters (or "gene families") in other species than collections of genes duplicated in their own genome.

Genomes, then, can evolve in two dimensions. The first dimension of evolution is through speciation. When two species diverge, the genes in their genomes diverge. The second dimension of evolution of genomes is through gene duplication. The bad news is that the combination of these two dimensions makes for some very complex patterns in tracing evolutionary history. If there were no duplication events in the history of life, then it would be very easy to trace the history of any genome. The good news is that this phenomenon could solve our rooting problem. If traces of a single gene family can be found in organisms of all three major groups of cells (Archaea, Eukarya, and Bacteria), then one can assume that the gene duplication event occurred in the common ancestor of all life on the planet, and a gene from one can be used to root genes in the others. The rooting of one gene family with a closely related gene family is called "paralog rooting" and was first used by Peter Gogarten at the University of Connecticut and Naoyuki Iwabe from Kyushu University. To date, the majority of duplicated genes in this category support putting the root at the base of the Bacteria. It appears from this approach that Archaea and Eukarya are more closely related to each other than either is to Bacteria.

Scientists who worry about evolution at the trunk of the tree of life are an odd bunch. Some of the first meetings to discuss building a tree of life were funded by the U.S. National Science Foundation and European funding agencies. The meetings were held to gain consensus on how a tree of life should be built and for scientists to present preliminary results on specific parts of it. At one meeting in Patras, Greece, the participants were asked to come up to a chalkboard and draw what they thought was the best-supported scenario for the ancient splits in the tree of life. An initial tree outline was drawn, and all attending were waiting for the session to start, when one of these odd scientists walked in, saw the tree on the board, and began to alter it with colored chalk. By the time he was finished, the chalkboard looked like the subway map for New York City, with crisscrosses indicating the large amount of horizontal gene transfer.

This story illustrates that the models we build as scientists are merely hypotheses to be scrutinized, tested, and retested. This is how science works.

Not surprisingly, then, when this seemingly overwhelming set of evidence for a "Bacteria first" hypothesis was settling in, some scientists challenged it. Their arguments were more than likely promulgated by the unquestionable similarity of Archaea and Bacteria. There are practically no scientists who go for the "Archaea first" hypothesis, so we can forget about that one. But the existence of some very bizarre, primitive-looking Bacteria brought up the question of whether there is a primitive Bacteria on which the tree can be rooted. In other words, maybe this primitive Bacteria is a member of a fourth domain and we simply haven't recognized it as such. So researchers developed other methods to seek out the root of the tree of life using three novel genome-based methods.

The first method looks at the distribution of short stretches of amino acids in proteins called "indels" (short for "insertions-deletions") that are present in some species and absent in others. The reasoning here is that these indels should be better tracers of evolution than changes in single amino acids in proteins, because it is assumed to be impossible to lose one and regain it in the exact same place. Although the results from this kind of analysis are somewhat controversial, they too indicate a root with Bacteria as the initial, most primitive lineage. The second novel way of rooting the tree of life is to use the characteristics of the genetic code to determine which species use more primitive elements of the code. The genetic code evolved as a means to enhance transfer of information between the genetic material and the proteins that a cell makes. In some cases, it is known that codons for certain amino acids arose before codons for others. By looking at the genome sequences of hundreds of Bacteria and Archaea, researchers were able to determine that all of the Bacteria used more primitive codons, indicating again that Bacteria is the odd man out.

The final genomic way to approach the tree of life concerns looking at networks of gene families. This approach takes advantage of how genes evolve as part of gene families and looks across various gene families to determine whether some genes or sequences of genes are present (or absent). After looking at over half of a million gene families, researchers concluded, once again, that some of Bacteria's gene families predate those of all of the other groups. This conclusion supports the idea that Bacteria comprise the original root in the tree of life.

Despite this overlapping and weighty evidence, there are still some researchers out there (including the researcher who used colored chalk to rain on everyone's tree of life party) who continue to argue that the root belongs somewhere else. These arguments primarily hinge on the anatomy of the cells involved, a topic to which we now turn.

Common Ancestors

One interesting way to look at the differences in cell characteristics is to consider the evolution of anatomical traits of cells. To take this step, we start with an idea of what a cell looked like before the two bifurcations or splits occurred in evolution. This hypothetical cell is affectionately called LUCA (for "last universal common ancestor") and is assumed to have some very basic traits of cells. This doesn't mean that LUCA arose out of nowhere in its full glory with those traits intact. Those traits and LUCA also had to evolve, with all of the lineages that sprang up on the way to LUCA going extinct. If LUCA came to the planet via an asteroid or a comet, this simply means that the evolution that led to LUCA occurred elsewhere in the universe.

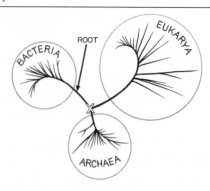

Figure 1.4. Phylogenetic tree showing the relationships of the three great domains of life—Archaea, Eukarya, and Bacteria—when the root is placed on the lineage leading to Bacteria. In this tree, the Eukarya and Archaea are each other's closest relatives. This is the tree that biologists accept as the most likely scenario for how the three domains of life are related to each other.

If we reject panspermia, however, then the evolutionary event to produce LUCA occurred on Earth. But it may not be so simple. Let's say Archaea and Eukarya share a trait we will call X, but Bacteria do not have it.

On the tree shown in Figure 1.4, we can point out where trait X arose, because we know the ancestral state of trait X was absent at the base (or root) of the tree. Now let's look at the second tree topology, where we use the classical Prokaryote hypothesis to describe the common ancestry of Bacteria and Archaea (Figure 1.5).

On this tree we can point out where trait X arose (see the two X's). For this tree, there are two ways to do this. The first is to suggest that trait X arose twice, once in Archaea, and once in Eukarya. In this scenario, the evolution of trait X would be what we call a convergence. For some reason, trait X was advantageous or even required for their lifestyle in both of these lineages and the trait was selected for. The second is to assume that the common ancestor of all life gained the trait and that it was then lost in Bacteria. This second

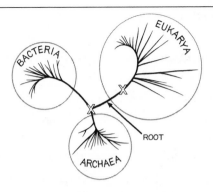

Figure 1.5. Phylogenetic tree showing the relationships of the three great domains of life—Archaea, Eukarya, and Bacteria—when the root is placed on the lineage leading to Eukarya. In this tree, the Bacteria and Archaea are each other's closest relatives and would form the group commonly known as Prokaryotes. This topology is now thought to be wrong.

scenario also requires two steps, but does not require a convergence. To make life easier, we need to choose one, and the one chosen is the tree that best explains all of the data. By choosing one, we can make sense out of the many cellular characteristics that have changed during the evolution of complex cells on the planet.

Using this procedure, one can search for the traits of cells that are common to all three major branches, assuming that these traits logically should also have been in LUCA. So what would LUCA look like? LUCA would have many traits even for such a primitive ancestor. Since all three domains of life use the same basic building blocks for heredity and for proteins, building blocks that are derived from nucleic acids and amino acids, these would logically be in LUCA. Sure enough, all three branches use the nucleic acids ATGC for DNA and AUGC for RNA, and they all three basically use the same twenty amino acids in their proteins.

Next, all three great lineages of cells use triplets to implement what is pretty much the same genetic code, and nearly the same system of redundancy. This would include using a polymerase (an enzyme that synthesizes DNA) to keep DNA double stranded, as well as a particular cadre of enzymes (enzymes to unwind the double helix, enzymes to stitch small fragments of DNA together, and enzymes to repair DNA when large numbers of mutations have accrued) to maintain DNA in the cell. LUCA would have had this machinery as well.

LUCA would also have adopted the same way of going from DNA to protein with RNA as intermediary that we see in all three cell lineages. In particular, LUCA would have had a complex cellular organelle called a ribosome, which works nearly the same way in all three lineages of cells and uses about a hundred proteins for cellular maintenance. These proteins would be essential for LUCA to obtain energy from fats, sugars, and other small molecules that it would encounter. In this sense, of the hundreds of different kinds of chemical reactions that can do these jobs, LUCA settled on very specific pathways that it channeled to all three of the cell lineages. Today the pathways for basic processes like sugar metabolism are essentially the same in all three lineages. LUCA would have also replicated in a very specific way by doubling its genetic material and splitting into two new daughter cells. Although eukaryotic cells have figured out more complex ways to do this, they still start with this basic plan of cell fissioning.

LUCA would also have had an outer layer or membrane to give it an internal integrity and protection from the external environment. Although the membranes of Eukarya, Archaea, and Bacteria differ significantly, they all have the same basic molecule as a starting point, and LUCA more than likely had it as well. Specifically, LUCA used what are called lipids to implement this barrier to the outside world. Lipids are long chains of carbon-hydrogen molecules that terminate with a special molecule that confers polarity to the chain, with one end that attracts water (or is "hydrophilic") and another end that repels water (is "hydrophobic"). Because these lipid chains have this love/hate relationship with water, they tend to line up in bilayers such that the water-repelling parts of the molecules are on the inside of the layer and the water-attracting parts are facing both the water-filled interior and the wet exterior. That is, the basic lipid bilayer structure in LUCA might have looked like the drawing in Figure 1.6.

Because LUCA had an interior and an exterior as a result of this lipid bilayer membrane, it would also have been capable of

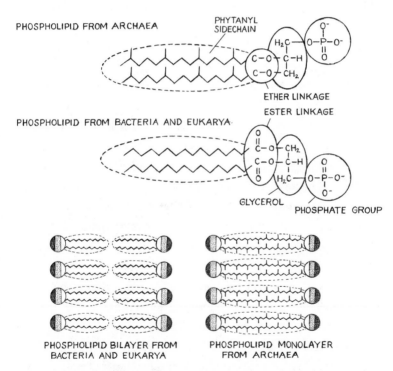

Figure 1.6. Schematic drawings showing the chemical structure of lipids in Archaea versus the lipids in Eukarya and Bacteria (*top*), and the arrangement of the lipid bilayer of Bacteria and Eukarya on the left and the lipid bilayer arrangement in Archaea on the right (*bottom*).

maintaining certain molecules on the inside that it liked and keeping out molecules that it didn't like. In fact, LUCA more than likely established a very specific set of conditions for its inner environment versus its outer environment. A higher concentration of sodium was maintained on the outside of LUCA than inside, and the converse was true for potassium. LUCA did this through the use of proteins called "ion pumps" embedded in the lipid bilayer. These small proteins literally pump sodium out of the cell and suck potassium into the cell. LUCA passed this system on to all three of the great lineages of cellular life. It turns out that even though we consider LUCA primitive, it actually was a rather complex entity.

The Rest of Life

The evolutionary biologist Stephen Jay Gould once pointed out that the age of dinosaurs, the age of mammals, and the age of man all pale in comparison with the age of bacteria, which basically has existed since about 3.5 billion years ago. LUCA set a course toward a biosphere composed mostly of microbes, beginning with the establishment of Bacteria, and with some branching of the tree of life, the other two great lineages of cells, Archaea and Eukarya (Figure 1.7). There are probably tens of millions of species of microbes that are alive today. If Bacteria and Archaea follow the same general rules of life on this planet, then 99.9 percent of all microbial species that have ever lived on this planet have gone extinct. This means that there have probably been tens of billions of bacterial species that lived at one time or another on this planet.

The characteristics of Bacteria are incredibly varied (Figure 1.8). They have found ways to be successful inhabitants of nearly every livable (and some nearly unlivable) place on this planet. When we think about humans, we consider ourselves fairly adaptable to the environment. Indeed humans exist on all seven continents in

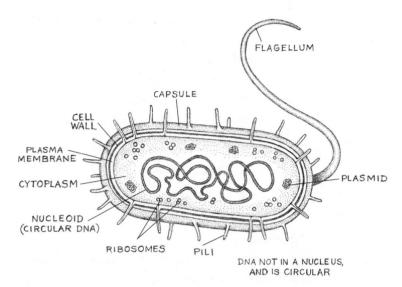

Figure 1.7. Stylized drawing of a typical bacterium, including its various structures.

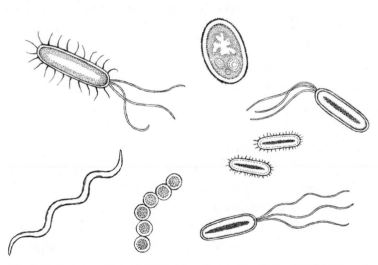

Figure 1.8. Drawing of some of the various forms that cells in the domain Bacteria take.

environments that range from extremely dry, subzero temperatures to environments with temperatures that soar to over 100° F and with very high humidity. Humans' adaptation to altitude—and to the various foods and liquids to which we expose ourselves—is well known. We are perhaps one of the most pliable mammals on the planet for adaptation to extreme environments. But our adaptability is nothing compared to the way microbes have evolved to succeed in the world.

Microbial Adaptation and Evolution

As groups of organisms started to fill open ecological niches and diverge from common ancestors, some of the ancestors' characteristics were preserved and others were changed as the DNA sequences of their genes evolved. For instance, one of the major first lineages of Bacteria to branch off from LUCA was the Chlorobacteria. The ultimate living fossils, Chlorobacteria are incredibly primitive organisms that have changed little over time and so have many of the traits that LUCA had. Other weird groups of Bacteria that branched off early in the evolutionary history of Bacteria, and hence can also be considered living fossils, include species like those from the genus *Deinococcus*. These species have evolved one of the best DNA repair systems in existence that allows them to live in environments with incredibly high radiation and still avoid the mutations in its DNA that would normally result. *Deinococcus* has even been used as a remedial agent in cleaning up radioactively contaminated spaces.

Another group that has some weird and primitive qualities is Cyanobacteria. Organisms in this phylum of Bacteria can photosynthesize like plants, because they have a set of enzymes that do it for them. Cyanobacteria and *Deinococcus* both move about by gliding around somewhat aimlessly on surfaces or in liquids, which suggests that LUCA moved in the same way. Later on Bacteria would evolve flagella for directed movement. Flagella are literally little whips that rotate to produce a swimming motion. (The origin of flagella is an interesting part of cell evolution, since all derived bacteria have them, as do Archaea and Eukarya.)

Archaea are a bit less varied than Bacteria, but they have also evolved to occupy incredibly diverse environments. There are several traits that Archaea share with each other that set them apart from other kinds of cells. They have incredibly stable flagella, have different proteins involved in DNA replication, and since diverging from Bacteria in the tree of life, have lost many genes from their genomes. By far the most telling traits of Archaea, though, can be found in the membrane. When membranes form, they can use building blocks made of molecules with a dextrorotatory (or "D") orientation, or they can use molecules with a levorotatory (or "L") orientation. The relationship of these building blocks is that of a mirror image, or chiral orientation. To convert one to the other is not just as simple as flipping them over. Your hands are the best example of chirality we can think of. Hold them up and you will notice they are mirror images of each other. You can't convert one to the other by simply flipping one. Try it; it doesn't work. LUCA's membrane was made of D building blocks, which were retained in Bacteria and Eukarya. Archaea, however, evolved the strange trait of utilizing L building blocks in the membrane.

In addition to this unique preference for L building blocks in the membrane (Archaea don't have this preference for L in other molecules such as DNA and proteins), Archaea also build lipids in their membranes by using a specific kind of linkage that is completely different than the linkages used by Bacteria and Eukarya. This means that the enzymes and proteins responsible for maintaining the archaeal membrane are very different from those that maintain bacterial and eukaryal membranes. Archaea also have flagella, but researchers in 2011 suggested that archaeal flagella are not the same type of flagella in eukaryotes and Bacteria; hence they are now called archaella to differentiate them. (Archaella and flagella most likely arose as the result of convergence.)

Eukarya have an obvious difference to Archaea and Bacteria: the nuclear membrane that surrounds the genetic material of the cell. The eukaryotic cell also has some extras that Archaea and Bacteria don't have: large organelles that carry out specific functions for the cell, specifically mitochondria, which occur in all eukaryotes, and chloroplasts, which in algae and green plants process light into food (Figure 1.9).

Mitochondria and chloroplasts offer insights into how Eukaryotes and Bacteria interact today. Why? Because at one time, both were bacterial cells that existed independently and moved freely in the environment. The common ancestor of all eukaryotes that lived more than one and a half billion years ago ate things by engulfing them. This allowed for relatively large items to exist on this common ancestor's menu, and led some Bacteria to be engulfed by hungry eukaryotes. This in turn led to a symbiotic interaction between

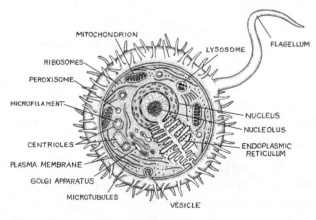

Figure 1.9. Schematic drawing of a stylized eukaryotic cell, showing its typical structures.

the Eukaryote and engulfed Bacteria groups that was cemented as the Bacteria became more and more dependent on the engulfed eukaryotic cell for simple cell function, and the eukaryotic cell gained a novel source of energy that allowed the cell to participate in aerobic respiration as well as gain other advantages. Eventually, the genome of the engulfed Bacteria was reduced so that the only genes in it today are those involved in the translation of proteins (ribosomal RNA and transfer RNA) and in the processing of energy. More than likely the bacterium that was engulfed started out with over a thousand genes. The genome of the modern-day mitochondria, by contrast, has just thirteen protein-encoding genes and another thirty or so RNA genes. The story with the chloroplast is similar. The common ancestor of plants more than likely engulfed an ancient species related to current-day Cyanobacteria, and this symbiotic relationship has been maintained throughout the evolution of plants as well as of some single-celled organisms. (Several groups of algae also have chloroplasts, but how these organelles evolved is a much more complex story.)

So now we have a sense of where microbes originated and how their genetics, traits, and functions have changed over time. But how does this history inform our understanding of how these organisms interact with us today?

CHAPTER 2

What Is a Microbiome?

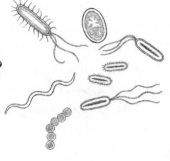

Who am I? This age-old question has been on human minds ever since we could think about thinking. But it wasn't until recently that the answer to that question was understood to be . . . mostly microbes. We have known for some time that microbes live in and on us. For example, who hasn't had a zit? The pus in a pimple is actually a microbial culture living under the surface of the skin. We have also known for a while that our intestines are full of microbes and that these microbes are essential for the proper digestion of certain foods. The pursuit of understanding how invading microbes cause disease, too, has been a major endeavor of microbiology for more than a century. In the past decade, however, as researchers have developed techniques that allow for a more precise and enlightening examination of what microbes live in and on us, we have learned that perhaps as many as 90 percent of the cells in a typical healthy human body are microbes, and that up to 3 percent of our body mass is made of microbial interlopers (so just go ahead and subtract that from your weight the next time you step on the scale). The perception that our bodies are made up entirely of human cells fighting off the occasional invader has been turned on its head, leading to a revolution in the way we think about microbes and their influence on our health.

Current estimates suggest that there are more than ten thousand different species of microbes in and on our bodies—approximately the same number as the species of birds that exist on the planet (Figure 2.1). Not all of these microbes are detrimental to human health. In fact, very few of them are harmful, and many have forged long and essential relationships with our bodies. As unbelievable as it may seem, our genomes have evolved to cope with and even cooperate with these microbial residents.

How to Detect Microbes

When one of our ancestors broke a bone while hunting or suffered an injury in battle, the source of discomfort was not terribly mysterious. The trauma was easily viewed as the cause of pain. Imagine though, living back then and waking up with a cough, runny nose, headache, stiff achy joints, or, worse yet, a terrible toothache. There would be no apparent trauma on which to blame the pain and discomfort. If it was a terrible toothache, you probably would have the tooth yanked out rather inelegantly and without anesthesia, but the root cause (no pun intended) would be as mysterious as most things in life were back then. This lack of knowledge about the source of the discomfort did not prevent our ancestors from dealing with some of the problems of infections from microbes, because all cultures developed folk remedies for a broad range of maladies. The problem was that our ancestors could not see the microbes. They knew there were some small organisms such as flies, mites, and maggots, but their natural vision was the limit of their knowledge about them.

The ancient Romans, plagued by seasonal bouts of devastating malaria (literally "bad air"), believed that foul smells rising out of the

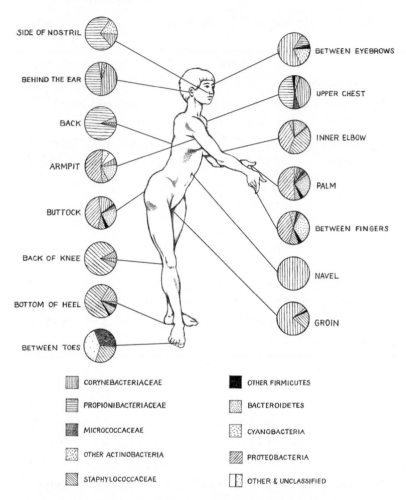

Figure 2.1. Diagram showing the human body and positions on the human body where microbial communities have been assayed in the Human Microbiome Project. The pie diagrams pointing to the body positions indicate the percentages of the kinds of bacteria shown at the bottom.

swamps produced the disease. Indian and Arab physicians hypothesized that microscopic parasites and mites caused various maladies in their patients. And Renaissance physicians such as Girolamo Fracastoro suggested that "fomites," spores of chemicals or particles, could cause infections. These ideas were all guesswork, however. It wasn't until humans made a concerted effort to actually look into the unseen world that they started to understand the relationship of microbes to the human condition.

One way of seeing unseen phenomena is to infer their existence with logical reasoning. Four hundred years ago, the process by which offspring of macroscopic, or visible animals and plants, arose from parents was relatively easy for natural historians and physicians to explain. But, as Aristotle pointed out, there was also a class of organisms where the connection between generations was not so obvious—they seemed to appear out of nowhere in a process called spontaneous generation. In 1668, Francesco Redi was the first to test the idea of spontaneous generation by placing pieces of meat in jars that were either covered with gauze or left open, noting that maggots appeared only when flies were able to access the meat. In the 1860s the French chemist Louis Pasteur showed that microorganisms likewise do not arise via spontaneous generation. If a nutrient broth was exposed to the air but no particles could settle into it (achieved by incubation in a flask with an S-shaped neck), the broth remained clear. But, Pasteur observed, if the neck was broken and dust particles could settle into the broth, it became cloudy, which was an indicator that microorganisms were growing in the solution.

Parallel to these experiments, other scientists were developing tools and methods to help them to see microorganisms. Starting in 1663, Robert Hooke was the first to use magnification to look at the fruiting bodies of fungi. This accomplishment was followed by Antonie van Leeuwenhoek's microscopic observations in 1675 of scrapings from his teeth and drops of pond water that he called

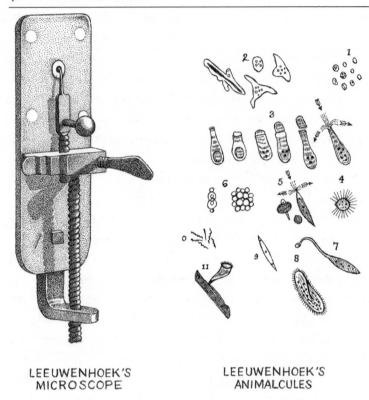

LEEUWENHOEK'S
MICROSCOPE

LEEUWENHOEK'S
ANIMALCULES

Figure 2.2. Drawing of Antonie van Leeuwenhoek's microscope (*left*) and some of the microbes he visualized with the microscope (*right*).

"animalcules" (Figure 2.2). These animalcules were the first bacteria and protozoans (single-celled eukaryotes) observed using microscopy.

Few may know that it was a German botanist named Ferdinand Cohn who actually started the field of bacteriology by looking at photosynthetic bacteria known as cyanobacteria. Through keen observation using the microscope, Cohn discovered that most bacteria could be categorized into four shapes—spheres, rods, threads, and spirals. Amazingly, these four categories are still in use nearly 150 years later. In 1872, Cohn was the first to describe *Bacillus subtilis,* an important species of bacterium that interacts with humans. More than likely his work influenced Louis Pasteur and Robert Koch, who are considered giants in the fields of medical microbiology. Pasteur's ideas about bacterial growth and contamination led to his testing of the bacterial version of spontaneous generation. Koch's work was more medical, and his ideas about infection in humans live on today with his famous postulates about infectious diseases in humans. Both Pasteur and Koch relied on being able to grow microorganisms in cultures in the lab. This was their way of visualizing bacteria and other microorganisms.

This is the point in the story when a couple of little-known names pop up in a big way: Martinus Beijerinck and Sergei Winogradsky. These two giants of microbial ecology didn't start out studying microbes. In fact, in his youth Winogradsky had hoped to become a classical pianist, and Beijerinck was a chemical engineer. Both, however, would turn to the world of the small and make impressive contributions to microbiology. Winogradsky specialized in stinky organisms and became famous for his work on sulfur bacteria. He also studied bacteria that are important in processing nitrogen in the environment. Martinus Beijerinck was the first to develop enrichment cultures for bacteria that allowed for isolated species to be grown in large amounts. Why grow bacteria in large, enriched amounts? The better with which to see them, whether through a microscope or by using some of the other inferential approaches around at the time.

Winogradsky developed his now-famous column in the late 1800s. This device was simple, but it led to some essential discoveries: the characterization of microbial trophic levels and a connection of organismal life to geochemical observations. It is easy and fun to make a Winogradsky-style column. Obtain a plastic or glass tube (a used two-liter plastic soda bottle will also do). Go to the nearest pond and obtain enough sediment from it to fill the vessel to about one-third volume. Next collect eggshells (which provide calcium carbonate) and crush them into small pieces. In addition to calcium carbonate, cellulose is needed for the column to work the way Winogradsky wanted it to, so shredded newspaper can be added. These materials should be mixed in with the pond or stream mud. There now should be a lot of mud, calcium carbonate, and cellulose in the bottom. Next add more of the pond or stream mud (without eggshells or newspaper), up to the two-thirds mark of the vessel. Top this with some of the pond water without mud, until the vessel is about 75 to 80 percent full. This arrangement of sediment and water creates a gradient of oxygen, with conditions at the top being highly aerobic (lots of oxygen) and those at the bottom being very anaerobic. Place the device in sunlight and allow it to culture for a month or two. Because specific bacteria in the pond mud and pond water will thrive in different concentrations of oxygen, the first set of reactions in the column will favor anaerobic bacteria in the lower layers of the column and aerobic bacteria in the top, until various sorts of microbial ghettos have been created in different parts of the column. Pretty soon the byproducts of the bacteria in the column will start to produce other gradients as well, in response to variations in say, sulfur concentrations or amounts of sunlight. These secondary gradients further subdivide the column, so that at the end of the culturing period the column will have stripes in it showing the preference of particular kinds of microbes for certain conditions. Winogradsky had created his own set of ecosystems, and this made it much easier for him to study the ecology of the microbes involved.

A culture of microbes can also be a soup of multiple kinds of microbes all swimming around together. Julius Petri invented in 1877 a dishlike structure with a gelatinous surface called agar and nutrients for growing microorganisms. The nutrients on the plate can be almost anything, but should be chosen to fit the nutrient needs of the microbe to be grown. To get single colonies of the microbes to grow, the concentration of bacteria is diluted using the century-old plate-spreading techniques. The spreader used is simply a fine metal loop that can be run through a flame to sterilize it. After spreading a gob of bacteria onto the plate in a large initial streak with oodles of bacteria, the spreader is sterilized, the plate turned 90 degrees, and streaks are made again overlapping the edge of the first streaks, which drags some of the bacteria from the first inoculation into a new section of the plate (Figure 2.3). The second set of streaks will now have fewer bacteria in them. The plate is rotated 90 degrees again and a third set of streaks is made. This last set of streaks usually does the trick, which is to isolate individual colonies of just one kind of bacteria.

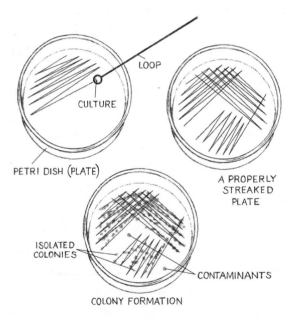

Figure 2.3. Petri dishes showing the streaking of microbes on the plate. The plate in the upper left shows the first set of streaks that are made. The plate is rotated 90 degrees and streaked again, followed by another 90-degree turn and streaking as shown in the upper right. The lower dish shows the final product that will result in single "isolated" colonies.

These individual colonies are then taken and grown in culture to create large amounts with which to experiment.

Another approach is to perform serial dilutions of a starting bacterial gob and to pour each dilution onto its own agar plate. As with the spreading technique, individual colonies can then be picked and further plated to isolate one microbe away from another, with the aim of creating large quantities of a single microbe species or strain for research.

Being able to culture microbes was a major advance for microbiologists. It allowed them to characterize microbes to a degree never before achieved. In medical microbiology, culture techniques were essential to the development of theories regarding how specific microbes cause particular diseases and to the formulation of an important cornerstone of medical microbiology devised by Robert Koch, a German physician, and published in 1890 as "Koch's postulates." These postulates provided a logical and feasible way to identify the causative agents of maladies. The first of these postulates suggests that the suspect microbe should be found in all individuals afflicted by a particular disease and not be present in healthy individuals. Koch's second postulate is that the culprit microbe must be able to be grown in pure culture. The third postulate is that the pure culture should cause the very same disease when introduced into another organism. And the final postulate is that one must be able to re-isolate the same culprit microbe from the secondarily infected organism. These four postulates worked well to identify problems caused by a single microorganism such as anthrax, cholera, and tuberculosis—diseases that Koch was able to link to specific types of bacteria. But when a community of microbes is the focus, or if the microbe needs special conditions or nutrients in order to be cultured, then the postulates fall short. While it may soon become possible to plate and culture a community of hundreds or maybe thousands of species of microbes, in the past it has been impossible. A new method called the iChip can speed the process up for soil bacteria. The iChip sorts single bacterial cells into small wells on a chip. The chip is then covered and placed back into the soil and allowed to culture for a while. In this way, researchers have been able to get cultures of impossible-to-grow microbes. This novel approach may speed up the process, but it will still be a tedious and time-consuming endeavor. So what's a microbiologist to do? Or even better, what did microbiologists do?

A Simple Solution

Microscopes, Petri dishes, and culture tubes—all of these have simple beginnings and are simple aids for the microbiologist. The problem of determining what is present in a sample of sputum or pond water, however, is a complex problem that requires much more sophisticated tools to solve. The challenge of finding out what is in a sample can be likened to taking a census of families in a neighborhood. The solution to taking a census is to knock on doors and ask the people who respond how many people are living in the house, as well as their names and various questions about their backgrounds. These bits of information are what we are after, too, when we try to characterize a sample. We need to "knock on doors" and ask which species are present, and, just as important, find out how much of each species is present in the sample. This sampling approach might seem easy in concept, but in reality it is pretty difficult to implement. The plating approaches we have mentioned could be used, but with hundreds of species present and different numbers of each species living in the sample, they can only give a fairly qualitative idea of what is there. Also, plating only works for those microbes that like to grow on agar plates with specific nutrients in them, and it turns out that most microbes in environmental and medical samples cannot be cultured with our current approaches. What is needed is a strategy that lines up the millions of microbes in a sample and asks,

"Hey—who are you? What species are you?" If one can characterize a large number of microbes in the line, then one can get a pretty
good idea of what is there.

The first intellectual leap in solving this problem was accomplished by the microbiologist Carl Woese, who, as we mentioned in
Chapter 1, realized that specific genes in the genomes of microbes
could be used to identify them and also determine their relationships to other microbes. For a gene to be useful as a diagnostic tool,
it needs to be present in every kind of cell. If a cell can't make proteins or molecules or copies of its DNA, then it will not live long.
So Woese and his colleague Gary Olsen at the University of Illinois
used a structure called a ribosome that all living cells have (because
it is responsible for translating genes into proteins). Ribosomes have
two subunits, the small subunit and the large subunit. These subunits are made of long stretches of RNA, which create complexes
with several proteins each. There is also a very short piece of RNA
embedded in the ribosome.

The three RNAs are called ribosomal RNAs, and they are each
identified by their size, with the shortest one called the 5S ribosomal
RNA (5S rRNA), the next largest designated the 16S ribosomal RNA,
and the largest subunit being the 23S ribosomal RNA. (The "S"
stands for Svedberg, a migration coefficient in a salt gradient, but for
our purposes it's most important to know that larger molecules have
bigger "S" values than smaller ones.) The 16S ribosomal RNA was
chosen as the bacterial identification target, because it is the most
conserved of the three types, thus allowing reliable use across very
different types of microbes.

The single-stranded ribosomal RNAs fold into what are called
stems and loops. The stems are made of base-paired parts of the
ribosomal RNA molecule. For the 16S rRNA, there are about thirty
noncontiguous little "ladders" with nonpaired loops interspersed
between them (Figure 2.4). The order of nucleotides in these

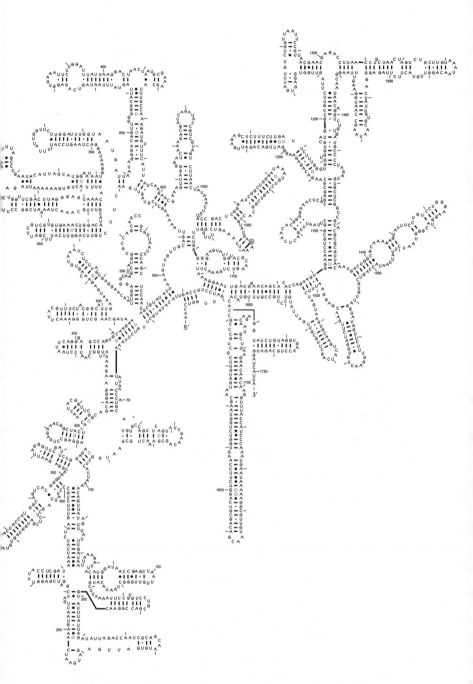

Figure 2.4. The secondary structure of bacterial 16S ribosomal RNA. The nucleotides in the RNA form pairs that are called "stems" and single-stranded regions called "loops."

loopy regions is what is used to determine relatedness and identity of microbes, so being able to sequence the gene that produces
the 16S ribosomal RNA—that is, identify which bases are present
and in what order—is critical. Luckily, the technology for DNA sequencing has been around since the 1970s, and now there are novel
technologies that have improved the ability to scale up and apply
this concept effectively.

To see how this works, let's look at a hypothetical sequence of
16S rRNA from four organisms stacked on top of each other, as in
Figure 2.5. Now let's try to represent the sequences in a different
way. Let's represent the four bases in DNA with different numbers
of lines. So a G will be represented with four very thin lines, an A
with three lines, a T with two lines, and finally a C with a single line
(Figure 2.6).

These kind of look like the barcodes on our food in grocery
stores, don't they? This is exactly what some researchers would
call this sequence information: a DNA barcode. In fact, if one set
up a scanner with this information, the four "species" would each
have an identifiably unique series of bands exactly like a barcode
one might find on a grocery product at the checkout. Now imagine
finding a bacterium that you think is new and after sequencing this
region of the 16S gene you see the following series of nucleotides:
GAATTACATT (Figure 2.7).

SPECIES 1   G A A T T A A A T A

SPECIES 2   G A A T T A C A T T

SPECIES 3   G A A T T A C A T C

SPECIES 4   G T T T T A T T T A

Figure 2.5. Short DNA sequences of four species showing the DNA barcoding
approach.

Figure 2.6. The four sequences in Figure 2.5 transformed into "barcode" lines.

Figure 2.7. The barcode of the "new" bacterium that also matches that of "species 2" in Figure 2.6.

Scan this with your eyes. Does it match any of the bar patterns above? A machine could scan these more easily, but eventually you and the machine will both come to the same conclusion: the sequence that you thought was a new species is actually the same as the species shown in the second line of Figure 2.6. We will return to this concept later in this chapter, because these DNA barcodes or DNA identifiers are at the heart of identifying which bacteria are present in a microbiome. The key to all of this barcoding is the easy-to-sequence 16S rDNA molecule, which exists in all living organisms on Earth. This gene is now perhaps the most sequenced gene in all of biology, with over two million sequences deposited into databases.

DNA sequences used in this way can help to identify an organism's position on the tree of life. If we compare a newly generated sequence (also called a query sequence) to sequences in a database, we will receive a list of sequences that range in similarity from zero to 100 percent. Those sequences with high similarity are usually either the same species or in the same genus. By setting the percent similarity used to categorize queries at varying values, we can obtain different levels of taxonomic information. Researchers will use a range of cutoffs to tell them what the higher level (less similarity) or species level (high similarity) composition might be. One other important aspect of doing comparisons with sequences is how we designate relative diversity. Microbiome specialists have suggested that there are two aspects of community composition that define diversity: alpha and beta diversity. Alpha diversity refers to the diversity of species within a single habitat or niche. There are different ways to measure alpha diversity, but in general, this aspect will be high if there are a lot of different species in the habitat and if they are all present in roughly equal numbers. Beta diversity, by contrast, is a comparative measure of diversity among different habitats. It will be high when the habitats all have large numbers of unique species present.

I See DNA Sequences

How does one determine the sequence of DNA? The nucleic acids are far too small to visualize, even with the most powerful microscopes. Although imaging technology has improved so that we can now see strands of DNA using powerful electron microscopes, this does not help us to translate what their sequences are.

Two people came up with solutions that can help us to read the code of a DNA sequence, and although both won the Nobel Prize, only one method, the one developed by Frederick Sanger, ended up being broadly used. Sanger, who passed away in 2013, is the only person to have won two Nobel Prizes for chemistry and one of only two scientists to win two Nobel Prizes in the same category. Sanger's "linear" way of thinking resulted in two of the most important scientific advances of the twentieth century: his other Nobel is for developing methods for discovering the sequence of amino acids in proteins.

The Sanger method of sequencing takes advantage of advances that had been made in *in vitro* methods for DNA replication. Kary Mullis had come up with the basic idea for polymerase chain reaction, or PCR, in the early 1980s. This method, still widely used in biology (and important enough to have garnered Mullis a Nobel as well), relies on a cyclical process whereby two strands are heated to melt the two strands apart from one another; the strands are cooled to allow short stretches of DNA called primers to bind to regions of the genome that they match; and then enzymes that transform single-stranded DNA into double-stranded DNA clamp onto those primers and fill in the second strand, thereby producing a new double-stranded piece of DNA. This process occurs along both strands of the DNA molecule that flank the region of interest, so as these three-step cycles are repeated, one can generate an exponentially large number of equal-sized fragments (Figure 2.8).

Modern Sanger sequencing uses a process that is similar to a PCR with two important differences. First, instead of including both primers, a sequencing reaction uses just one, so that the resulting products are all from the same strand. Second, in addition to mixing in regular nucleotides to be used as building blocks during the copying phase, special nucleotides are added. These nucleotides, called "dideoxynucleotides," lack a hydroxyl group and so terminate the strand at whatever position they enter the reaction, creating a strand of a particular length. In the early days of DNA sequencing, the dideoxynucleotides were labeled with radioactivity and visualized

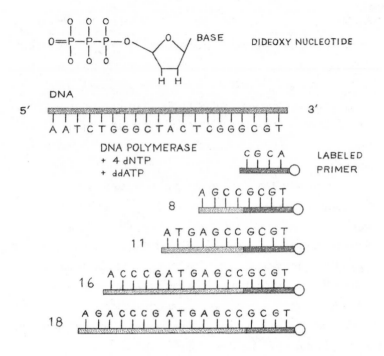

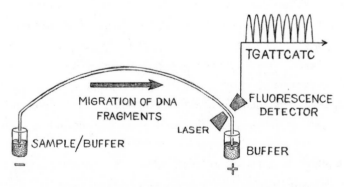

Figure 2.8. The Sanger DNA sequencing reaction. The diagram on the top shows the chemistry of the reaction; the diagram on the bottom is a schematic showing how a Sanger sequencing machine works.

using x-ray film; in modern sequencing, they have fluorescent dyes attached to them, with different colors depending on which kind of base they are—A's have a green dye; T's a red; C's, blue; and G's, yellow. After repeated cycles, just as described for PCR, the reaction tube will contain a population of molecules with members that end at every possible base along the sequence. The molecules are then separated by size, by pulling them through a very fine tube containing a polymer that slows down their travel so that the smallest fragments get through first, with sequentially larger fragments taking longer to make this journey. If a laser is applied to the molecules as they reach the "finish line," we can read the sequence. If the first color we see cross the line is blue, then the first base after the primer was a C; if the next color is red, then the next base was a T; and so on.

The Match Game

The tragic events of September 11, 2001, resulted in the deaths of 2,792 people in the two World Trade Center buildings, most of whose remains were dispersed across the disaster site in what would become 19,906 separate samples. Forensic investigators faced with the task of identifying the remains turned to DNA sequence analysis. DNA was isolated from the remains, many of which were bone fragments, and then characterized for various genetic markers. The DNA sequences from the remains in and of themselves were somewhat useless in the identification process—what was needed was a database of sequences with known associations. So the relatives of victims of the attacks were asked to donate blood for DNA analysis to build up a database that could be compared to the remains. To date, the remains of over eight hundred individuals killed by the attacks have been identified using this approach.

More akin to our microbial community problem are the national DNA databases set up by many countries to aid in the identification of criminals using DNA sequences. In the United States, the FBI has established a database called Combined DNA Index System, or CODIS, that stores the DNA sequences of anyone who has been convicted of a crime. CODIS has nearly 11 million entries in it. This database is used in criminal investigations to determine if a perpetrator of a crime who has left some DNA at a crime scene is in the database. The U.S. Armed Forces has instituted a similar identification system. In addition to fingerprinting every inductee into the armed forces for identification purposes, the U.S. military takes a blood sample for DNA analysis and deposits the information in the Defense Enrollment Eligibility Reporting System, or DEERS. There are currently about 1 million entries in this database.

Since DNA was first used to identify microbes, researchers have also stored DNA sequence information in databases. At first the sequences were simply stored in the national repository for DNA sequences at the National Library for Medicine in the National Center for Biotechnology Information (NCBI) at the National Institutes of Health (NIH), a site affectionately called GenBank. This DNA sequence repository was formed in 1988 by Congressional order to serve as a repository for DNA sequences from any and all genes and from any and all organisms. The NCBI repository is accessible to the public and serves as an invaluable resource for genetic research all over the globe. Currently it has well over two petabases of sequence in it—that's 2,000,000,000,000,000 G's, A's, T's, and C's—from over a quarter million different species. Some species are much better represented than others. Not surprisingly, sequences from thousands of *Homo sapiens* individuals make up a large proportion of the database.

Microbiologists recognized early on that having a database of bacterial DNA sequences would be an invaluable resource for identifying bacteria. In 1989 the National Science Foundation funded the Ribosomal Database Project (RDP), which in 1992 went online with the sequences of 471 16S rDNA sequences as the core of the database. The Argonne National Laboratory hosted this first database, which was moved to the University of Illinois in 1995 and then to Michigan State University in 1998. As of early 2015 there were nearly three million 16S rDNA sequences in the RDP. Since each 16S sequence is nearly two thousand bases long, this means that nearly 6 billion bases for this single molecule have been cataloged in this database.

Many of the sequences in this database are not named in a taxonomic sense, but some of them do represent branches of the microbial tree of life that are major players in microbial communities. Remember that there are only about eight thousand named species of Bacteria and Archaea, but millions upon millions of species out there. By storing the sequences of unnamed but phylogenetically important species in the RDP, researchers can at least use these unnamed species as placeholders for important kinds of organisms in, on, and around our bodies.

Naming All the Needles in a Haystack

When microbiologists first started to look at microbes in communities (whether they were in body cavities or in environmental samples), the process of identifying what was there was incredibly labor intensive and time consuming. After obtaining a sample, such as a fecal sample from a person or a water sample from a stream, researchers would laboriously plate out the bacteria from that sample on a Petri plate. Whatever grew on the Petri plate was then isolated and cultured, and then each culture was characterized and designated as a particular species based on its biochemical characteristics. This process was very difficult to accomplish if there were many different kinds of microbes present in the original sample, but, more important, any microbe that didn't grow on the original plate would be missed, because if it didn't make it past the first culturing screen, it couldn't be isolated and tested. Yet the effort was still worth it: initially, this process of identifying microbes led to a lot of very important discoveries involving microbes that live in and on our bodies as well as microbes of economic importance. After a little over a century of characterizing and naming bacteria this way, almost eight thousand different species had been characterized and named. Moreover, many of the bacteria discovered were pathogenic, and having cultures of them helped advance research into microbes that make us sick.

But although this work produced a collection of microbes important to human health and industry, it was not an accurate representation of the actual diversity of microbes in the natural world. In fact, it wasn't until microbiologists started to recognize that there were species of microbes they could not culture that an amazing number of microbes—and more important, an amazing number of kinds of microbes—emerged. Eschewing the tedious plating and culturing techniques, the microbiologist Norman Pace, then at Indiana University, reasoned as early as 1985 that any sample, whether of dirt, pond water, or a bodily fluid, would contain all of the DNA of all of the microbes in the sample. The trick would be to separate out the individual 16S rDNA identifiers in the sample and to identify them.

This situation brings one back to the initial problem faced by the classical microbiologists, who would solve the problem with brute force by plating and culturing. But Pace accomplished this goal in the molecular biology lab by cloning the entire mix of DNA in the sample (16S plus the thousands of other genes present) into bacterial cloning vectors and then using selective techniques to recover from the mixture only those bacteria that contained the vectors with 16S rDNA. This approach was a bit quicker than the plate and culture approaches, and it was more comprehensive: it yielded 16S rDNA sequences of organisms that could not be cultured. Sequencing these clones and characterizing the results gave a good idea of what microbes were living in the sample. This approach is known as DNA shotgun sequencing, because the dataset of results resembles the scatter of a shotgun blast.

Pace and his colleagues published the first such shotgun analysis in 1990, and this was followed by a flurry of studies that examined a wide variety of microbial environments, mostly in an ecological context. Pace and his colleagues first collected a sample of ocean water from the Pacific Ocean at the ALOHA Global Ocean Flux Study site (at 22°45′N, 158°00′W) from aboard the RIV Moana Wave. Once Pace obtained the shotgun sequences, he and his colleagues used a simple genealogical tree-building method to determine which were the closest known microbial relatives to the clone sequences he had found. This is tantamount to using the CODIS database to identify a forensic sample. When Pace and his colleagues used this technique they were able to identify many of the unknowns in their sample, though many of them were bizarre and represented totally new higher categories of microbes.

Other researchers then recognized that they could easily simplify the shotgun-cloning approach by first isolating the 16S rDNA away from all of the other genes in the mix. They determined that if one took an environmental sample or a mixture of millions of different microbes and amplified only the 16S rDNA genes using PCR, the vast majority of the DNA present would be the mixture of 16S rDNA from the various microbes in the sample. This meant that the researchers could skip the tedious 16S rDNA identification process.

They then reasoned that they could take the 16S rDNA mix and clone it into plasmid cloning vectors, which do a great job of individualizing and further amplifying segments of DNA. They simply took the mixture of 16S rDNA fragments made from PCR, cloned them into vectors, and picked hundreds of the clones for sequencing. The approach was not only quicker but also facilitated the acquisition of more 16S rDNA clones, which increased the sample size of new microbes. In back-to-back papers in *Nature* in 1990, two groups showed that the PCR approach coupled with clone sequencing had identified a large number of previously uncultured microbes in the Sargasso Sea and in hot-spring microbial communities. These first studies looked at only tens of clones. But in 2003, on the heels of sequencing the first human genome, Craig Venter undertook a more thorough analysis of Sargasso Sea communities. This analysis indicated that more than two thousand different species were present in most of the communities he had sampled from this part of the Atlantic Ocean, and of these, nearly 150 were types of bacteria previously unknown to microbiologists.

It is difficult to pinpoint the very first microbiome or human microbial community study, but one of the first of which we are aware, which focused on the oral cavity, was undertaken in 1994 just after Norman Pace's pioneering work on environmental samples. By the early 2000s, the use of DNA sequencing to characterize microbial communities in and on humans had taken off. In 2001, Stanley Falkow and David Relman, two medical microbiologists, made the bold proposal that a second human genome project be initiated in order to sequence and catalog the thousands of microbes that inhabit the human body. Specifically, they called for the inventory and sequencing of the thousands of bacteria associated with our bodies at four main sites of colonization: the mouth, gut, vagina, and skin. That same year, Joshua Lederberg coined the term microbiome to describe the assemblage of microbes associated with the human body. It was suggested that "characterization of the microbiome would be accomplished through random shotgun sequencing procedures, targeted large-insert clone sequencing." Although these scientists were right about characterization of the microbiome, the invention of some techniques was required before microbiomes would become as fascinating as they are today.

It's Generational

There were at least two important outcomes of these early studies using shotgun Sanger-style sequencing. First, these early microbiome and environmental studies demonstrated that, in principle, researchers could gain access to census information for microbial environments. Second, the studies showed that science had indeed been missing major lineages of microbes that could only be found using approaches that don't rely on culturing. For example, let's consider a favorite kind of microbe—the fungi. Fungi are a large group of nearly 100,000 named and described species that include mushrooms and yeast. The major groups of fungi fall into two phyla—Basidiomycota and Ascomycota—which make up around 90 percent (90,000) of the known species. According to Meredith Blackwell, there are around ten other fungal phyla and there may be as many as five or so million species in the overall group Fungi. In a 2011 study of pond water on the grounds of the University of Exeter, a faculty member, Thomas Richards, and his colleagues found an incredibly diverse and taxon-rich new fungal lineage. They dubbed the group Cryptomycota, or the "hidden fungi," because this group had remained undiscovered until Richards and his colleagues went pond dipping. This is a rather startling discovery, and a bit like discovering a new group of animals unlike any of the other phyla of animals we have on the planet.

Bacteria and Archaea have had similar amazing discoveries. The trees shown in Figure 2.9 show what bacterial phylogeny looked like in 1987, before the culture-free approaches were invented, as well as in 1997, 2003, and 2004. Note that all of the bacteria in 1987 were cultured strains. In 1997, when the culture-free approaches began, about one quarter of the divisions in bacteria were uncultured. In

KNOWN BACTERIAL PHYLOGENETIC DIVISIONS

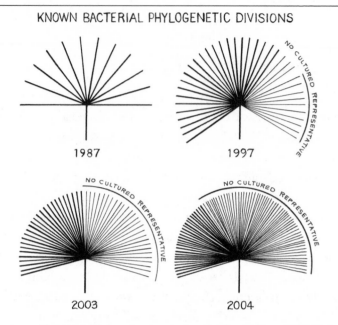

Figure 2.9. Phylogenetic trees showing the increase in number of identified bacterial species as techniques for viewing and identifying them improved.

2003, more than half were uncultured, and the number of major divisions of bacteria had almost quadrupled. Also note that during the four years from 2003 to 2007 even more divisions had been discovered, and five of six of those are uncultured.

Although these methods of sequencing 16S clones from samples catapulted our knowledge of microbial diversity forward, researchers would still miss a lot of information unless great effort was placed into the project (such as with Craig Venter and the Sargasso Sea analysis). This missing information is caused by a phenomenon known as rarefaction. Microbiologists use what is called a rarefaction curve; the statistics associated with the curve help determine if they have done a good job at finding as many different kinds of microbes as possible in these kinds of studies. The math of rarefaction is a bit

complicated, but fortunately the curves are pretty self-explanatory.
Some examples of curves to illustrate how rarefaction is important
are shown in Figure 2.10. In the graphs we show what a rarefaction
curve for a poorly surveyed community, and a well-surveyed community, would look like. For each graph, the x axis indicates the
number of sequences that were generated from the original sample,
and the y axis tells how many species were found for different sample sizes. For the poorly sampled community, we see that we have
not yet found all of the different species in the community, because
when we are at the maximum number of sequences, the curve has
still not topped off. By contrast, the community on the right is a
well-sampled one. This is true because after only about a quarter of
the total sequences have been characterized, there is no real increase
in the number of new species discovered. Most rarefaction curves
for studies done with the shotgun approach would look something
like the curve on the left, and hence would require a lot more sequencing to fully characterize the diversity in a community. This is

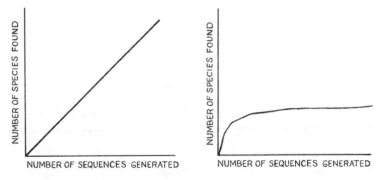

Figure 2.10. Graphs showing how rarefaction works. The diagram on the left
shows a sample where not all of the species in the sample can be found even
after extensive sequencing. The diagram on the right shows a sample where
the true estimate of the distribution of species has been estimated after much
less sequencing.

not to say that studies done then are inaccurate or useless. Rather, their only flaw is that very rare species oftentimes were missed.

Note also in Figure 2.9's tree diagrams that there was a large increase in the discovery of new bacterial phylogenetic divisions between 2003 and 2007. Getting to the 2007 level of discovery required a quantum leap: new, ingenious methods for DNA sequencing. These methods, which were invented around 2004, are called next-generation sequencing (NGS) approaches, and they increased the amount of sequence that could be obtained in any biological experiment by several orders of magnitude. The solution lay in miniaturizing the sequencing reactions and hence allowing for more sequences to be generated in a small space.

There are many NGS approaches being developed that promise spectacular advances in technology, and two have been used extensively in characterizing the microbiome. The Roche 454 platform was the first to be used to examine community structure in microbial systems. Sanger sequencing reactions had traditionally been done in a microtitre plate about two inches by four inches that could accommodate only a maximum of 384 reactions. The 454 approach did away with plates and tubes for doing reactions. Instead each DNA molecule to be sequenced is anchored onto microscopic beads, which are each placed in an emulsion that encloses each bead within its own pool of reactants and prevents the reactions from interfering with each other. The reactions are processed en masse and then placed onto a chip with little depressions in it, one for each bead. The 454 chip is at most half the size of the Sanger plate, yet it has 400,000 depressions in it, making it more than a thousand times more powerful than the Sanger methods.

Once the beads have been distributed into the wells on the chip, the chip is placed into a machine that is essentially a microscopic camera. While it is held firmly in a stable position, reactants are poured over the chip, with each reaction indicating the presence of a different base (G, A, T, or C). If a G is being washed across the chip and a particular well has a piece of DNA that should react with G, then a small flash of light occurs. Pictures of the chip are taken to detect the flashing as a result of the reactions, and the timing of the reactions and the adding of the bases are kept track of by digitally "filming" the chip. In addition, the position of each of the 400,000 depressions in the chip is tracked. Ultimately, if all goes well, the researcher will have 400,000 sequences representing 400,000 microbes from a sample. In terms of the next-generation sequencing framework, this result would mean that the run generated 400,000 "reads," where a read is information regarding a single sequence. But there are clever ways of tagging reactions so that one can put several different communities on a single chip. For instance, if one had four samples to evaluate, one would tag the products of the four samples with different markers and sort them out after the chip has been run. This would thus leave us with each community sample having 100,000 reads.

Another platform developed in the past five years that has been used to characterize microbial communities is called the Illumina platform. Although the miniaturization principle is also used in the Illumina platform, instead of a chip, a flow cell is used. This platform also uses a different chemical approach, which allows for the generation of up to thirty billion reads. The Illumina reads are a bit shorter than the 454 reads, but the sheer number of reads is stunning, making for five to six orders of magnitude more reads. In addition, because microbiologists have localized pretty nicely where in the 16S rRNA most of the variation occurs, small fragments are not a problem.

Even More "-omes"

If you thought that your body harbors a lot of living microbes, then, as they say, "You ain't seen nothing yet"—for the ante is upped by orders of magnitude when viruses are included. If, as Stephen Gould once said, our planet has existed for 3.5 billion years in the age of bacteria, then bacteria have always lived in the age of viruses. Remember that these simple entities are not considered life by most scientists. But there are so many of them coursing around inside of bacteria and us that they demand our attention.

The characterization of the viromes of individuals will likely be an important step in addressing many human health issues. There are many problems that researchers and clinicians confront, however, when thinking about characterizing viromes. The first and foremost problem is that the virome is made up of a much more dynamic and changing composition than the microbiome. One's diet, geography, age, genetic background, and lifestyle would all be involved, as well as the time of year when the viromes would be characterized. Another problem is a technical one that concerns the diversity of viruses that inhabit our bodies. As discussed previously, viruses come in many sizes, shapes, and kinds. Some have circular chromosomes that carry their genes. Some have their genes broken up into short linear chunks of genes or even single genes. Some use RNA as a means to store and replicate their genes, and others employ DNA. Some use double-stranded DNA or RNA and others use the more streamlined single-stranded forms of these nucleic acids. All these aspects of viral genomes make developing a universal marker for studying viral community assemblages such as microbiomes very difficult. (Remember that, by contrast, all living microbes use double-stranded DNA to replicate and store their genetic material.) More important, they all share many of the same genes that they use to replicate DNA and to make proteins from their genes. Whereas scientists can characterize microbiomes simply by looking for the microbial genes that code for ribosomal RNAs, the diversity of viruses' size, shape, structure, and targets and the fact that no single gene is found in all viral genomes make it difficult to devise similar approaches for viral community assessment.

One approach that has been tried involves a protein that exists in a large number of viral species' genomes called the RNA-dependent RNA polymerase (RdRP) gene. This gene has been used to identify various major kinds of viruses much like the 16S rDNA gene has been used as an identifier for Bacteria and Archaea. For the most part, though, the identification of viruses in the human body using high-throughput approaches has to follow the route of DNA fingerprint databases. In other words, researchers need to identify and sequence all viruses in humans in detail and place these sequences in a database. If one has a reliable database, then detecting which viruses are in a sample becomes technically feasible. Even so, detection of novel viruses that may be involved in emergent and new infectious diseases will be relatively difficult unless researchers concentrate on discovering and characterizing new kinds of viruses much like museum curators discover, archive, and name new species of antelopes and new species of wasps.

Those readers who have stayed up all night with a feverish young child may be interested to learn of pathbreaking research on fevers that has resulted from just this sort of viral cataloging. By sampling and studying the viromes of children with fevers and without, scientists have discovered that children with fevers appear to have nasal samples with 1.5 to 5 times more viruses than kids without fever. Kids with fever also have more and broader types of viruses present, not just more viral particles of one kind. A final interesting aspect of studying the virome of feverish children is that the big differences occur when looking at the child's blood plasma. All other samples taken, such as urine or feces, do not show the large changes in virome between children with fever as compared with children without fever. This is more than likely because plasma is the best route to spread an infection in the human body, and there is no better place to find viruses than on the highways they traverse to cause infection.

All these results show a connection between fever and the virome that, once deciphered, may lead to more effective ways to treat fever—and just as important, remedies to avoid. In particular, if a child has a fever and the virome can be characterized and shown to have a larger-than-expected number of viruses with lots of diversity, then treatment with an antimicrobial should be avoided, because the infection causing the fever is more than likely the result of a viral infection, and antimicrobial treatment would both be ineffective and disrupt the microbial assemblages. Indeed, antimicrobial therapy should be avoided in most cases of fever and avoided at all costs when the cause of an infection is viral.

Other tiny organisms also live in and on us. Infections such as jungle rot (caused by various organisms, including mycobacteria), athlete's foot (a fungal infection), Guinea worm (caused by drinking water infected with parasite-ridden tiny crustaceans called copepods), or giardiasis (probably the most common pathogenic parasitic infection in humans worldwide, caused by the flagellate protozoan *Giardia lamblia*) can be extremely unpleasant. Compared with bacteria and other microbes, however, these kinds of organisms are by far lone wolves when it comes to influencing human health. For instance, although ten thousand species of bacteria may reside in and on us, there are just ten to a hundred species of these eukaryotic organisms that are interlopers on and in our bodies.

Human bodies are not a single simple ecosystem. In fact, seemingly closely located parts of the body can have radically different environments and hence ecologies. The ecologies on and in bodies can also vary over time. After a workout the underarm is wet, hot, and salty. Take a shower and pat dry, and the environment changes about as drastically as if we went from the jungles of Panama to a freshly cleaned house in Detroit, Michigan. As we age, too, the microbes with which we associate change, so the age of people who are sampled becomes important. In addition, there will be differences between females and males, especially in those areas of the body that are anatomically different between the two sexes. It also makes intuitive sense that the microbiomes of people in different geographic areas will be different. And then there are sampling differences between people who are very healthy and people who might be ill at the time of sampling.

Next-generation sequencing approaches have changed the game for the discovery and characterization of microbial communities. A typical microbiome study in 2013, using the next-generation sequencing approaches, will have between 1,500 and 10,000 reads per sample. In fact, Relman and Falkow's charge to medical microbiologists is now being realized because of these technical advances. So much momentum has been gained by next-generation sequencing approaches that in 2009 the National Institutes of Health initiated

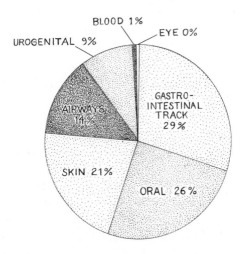

Figure 2.11. The approximate distribution of microbes on various parts of the human body.

the Human Microbiome Project. Obtaining the microbiome information from the human body might seem simple at first thought. Just swab parts of the body and go for it with the next-generation sequencing approaches and get pie charts like the one in Figure 2.11. But the NIH and the originators of the Human Microbiome Project (HMP) first needed to address some complicating factors before throwing 150 million dollars in slated funding at the project. To do so, they first sampled the body parts, crevices, and surfaces of 250 volunteers (half male and half female) in order to maximize the discovery of new species. The pie diagram shows the number of species potentially on or around the different parts of our bodies. These data guided the next step: choosing eighteen specific sampling sites in and on the 250 volunteers. After this initial group was sampled, the complete project was begun. The HMP is now in full swing, and the results from it, as well as from related projects by other researchers, are accumulating at a spectacular rate.

CHAPTER 3

What Is On and Around Us?

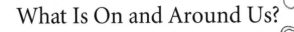

At the end of the movie *Dr. Strangelove,* the main character, while discussing how to survive the impending nuclear holocaust, cannot control his right arm from shooting out into stiff-armed salutes to his former leader. At one point, he even uses his left hand to beat his right hand into submission. This humorous cinematic depiction of the biblical quote "don't let your left hand know what your right hand is doing" might just have a microbial equivalent. By analyzing the microbiome of left and right hands of several undergraduate students at the University of Colorado, Rob Knight and his colleagues uncovered some very Strangelovian (or Matthewian?) patterns.

The purpose of the study was to see if microbiomes on right hands were any different than those on left hands, but there were two additional goals. First, the researchers were able to sample males and females in relatively equal numbers, so they could compare the microbial communities between male and female hands. In addition, a smaller sampling was undertaken to look at the effects of hand washing on microbial diversity. There were several rather stunning results of the study. First and foremost, the researchers discovered a large diversity of microbes (on average over 150 species of bacteria) residing on each of the hands of these undergraduates. Even though all of these students lived in the Boulder, Colorado area, a bucolic community located more than six thousand feet above sea level, more than 4,500 different kinds of bacteria overall were found on the students sampled.

Examining the other variables in the study—hand washing, male versus female subjects, and handedness—also gave important and rather surprising results. One might suppose, for instance, that students who washed their hands would have smaller communities of bacteria on them. (After all, isn't this the purpose of hand washing?) But it turns out that time since last hand washing was not correlated with diversity and, in fact, the degree of diversity is pretty much the same regardless of time since washing. What did change over time were the kinds of bacteria found on the hand since the last washing, implying that there is a kind of succession of bacteria that colonize a clean hand. Certain types of bacteria jump on at first and establish themselves, and then a little later others take over.

The differences between sexes were unexpected as well. One might guess that since most men are notoriously less hygienic than most women, men would in general have more microbes on their hands. But in fact the reverse is true: females have more microbial diversity on their hands than males. The reason may have to do with differences in the microecology of the skin of men and of women. In general, men and women differ in the acidity of their skin, which may allow for different microbes to live on the skin of women versus men. In addition, men tend to sweat more than women, and this tendency might also decrease the number and therefore the diversity of microbial species on the palms of men.

But the strangest of all results from the study is about handedness, and it is so Strangelovian that it tells an impressive story about ecology. The researchers found that only 17 percent of the microbes on the dominant hand were also found on the nondominant hand. Truly, the left hand does not know what the right is doing.

Characterizing microbial diversity is difficult because of the sheer numbers of organisms. But progress can and has been made in detailing the millions of bacterial species with which we are as-sociated—we just haven't gotten around to naming all of them yet. Many interesting questions about the association of microbes with humans can now be addressed with the next-generation sequencing approaches that were invented in the past five years, and there will be many more questions, and approaches, to come.

Skin Game

As we mentioned earlier, the NIH has developed a large-scale, $150 million effort called the Human Microbiome Project (HMP) as an important step in understanding our microbiome. A subproject within HMP is the Skin Microbiome Project, which examines the microbiota of the largest organ of the human body.

The skin's structure is an evolutionary marvel: it has evolved to be a barrier to the outside while keeping organs and tissues inside, and it does this extremely efficiently. Organisms such as bacteria, viruses, and fungi can land on the skin, but have a difficult time pen-etrating unless there is a cut or an orifice where they landed. Indeed skin is the body's first line of defense against infection. The environ-ment a microbe encounters on our skin is going to be acidic. Most microbes have only narrow ranges of tolerance to pH, so if the skin has an acidity to it that is different from what a bacterium is used to, it will not fare well. The skin is also a bit cooler than other parts of the body (such as the oral cavity or the intestines) and it is very dry, unlike other places on the body where mucous layers exist. So any bacterium that alights on the skin and starts to grow is one that has adapted to the acidity, temperature, and moisture conditions of that area of skin.

From a microbe's perspective, the topography of the skin's surface looks as varied as a relief map of the state of Colorado. Some parts are dry and cool, but there are also warm and moist areas such as the groin, underarms, and the folds of the inner part of the buttocks. There are rolling hills on the skin—more as one ages—as well as deep, forbidding pits near the surface (the sweat glands) and towering treelike structures, or hair follicles. Further dissection of the skin's microhabitats requires learning what constitutes the cellular level (Figure 3.1). The top layer is made up of specialized differentiated cells called keratinocytes. These cells are chock full of proteins, including keratin, a tough, fibrous protein that makes these cells impermeable to outside elements. These keratinocytes have a specific name—squames—and they are continually being renewed as the skin regenerates. They are like tightly mortared bricks in a building. The topology of the skin is varied, too, because of the way

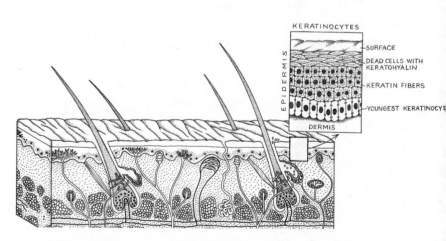

Figure 3.1. A cross-section of human skin, with a close-up view of keratinocytes.

the skin cells articulate and depending on the region of the body. There are probably more different kinds of habitats for microbes on the surface of the body than there are for humans all over the planet.

By taking swabs from nearly every imaginable region on the exterior of the human body, researchers in the Human Microbiome Project have made some incredibly interesting observations about the diversity of the microbes living on us. Before we report these results, learning a little anatomical nomenclature will be helpful. For instance, the region behind the ears is the retroauricular crease. (Imagine a mother telling a child, "Be sure to wash your retroauricular crease!") The part of the arm where a crevice forms when it is bent is an antecubital fossa ("Put some antecubital fossa grease into it!"). But our favorite is the umbilicus, more commonly known as the belly button.

There are basically four general kinds of bacteria living on the skin: Actinobacteria, Firmicutes, Bacteroidetes, and Proteobacteria, each of which is what taxonomists call a phylum, a large category used to classify organisms, at the same level of Vertebrata or Arthropoda. Although there are just four phyla of microbes living on our skin, there are numerous types of each.

Because the formal description of species is challenging for Bacteria, one of the terms that microbiome researchers use is called a phylotype, which is a nonspecific way of giving something a provisional name. In this approach, an arbitrary cutoff, usually the percentage difference in the barcoding sequence between isolates, is used to determine that one organism is in a specific phylotype and another is not. So, for instance, if we have sequenced the barcode region for two microbes (A, B), we can compare these sequences to that of a reference species (R). If the percentage difference between A and R, and B and R, were both below a specific cutoff, then they would be considered the same phylotype as the reference species.

If the distances are above the phylotype cutoff used, then they are considered different phylotypes. If the information on the difference between A and B is included, it is also possible to determine if A and B are members of the same phylotype. Although some researchers prefer to give names to their study subjects, the phylotype concept has been an important first step toward understanding the diversity of microbiomes. The phylotypes in skin microbiome studies attempt to get at what a species is, but since many of the microbes living on our skin have not been named, or cultured for that matter, species names aren't used.

The skin microbiome project has revealed that the phylotypes discovered fall into three main categories, based on the type of skin surface where they are found, but not necessarily related to their location on the body. The first category corresponds to oily or waxy skin areas, which are termed sebaceous regions. The diversity is lowest in these regions, but even here there are several phylotypes present. The back is considered a sebaceous surface, and it has around twenty phylotypes, whereas the forehead, also considered a sebaceous surface, has only about six. Most of these phylotypes of bacteria found in these skin regions are in the genus *Propionibacterium*. This genus has been studied a lot, and species in it are unique in their ability to produce proprionic acid. They are also lipid lovers, a preference that allows them to colonize pilosebaceous units that are positioned all over the skin. This group of bacteria is involved in acne—that scourge of the teen years.

On our moist skin areas the major phylotypes are in the genera *Staphylococcus* (a member of the Firmicutes phylum) and *Corynebacterium* (a member of the Actinobacteria phylum). Both of these bacteria take advantage of components like urea in sweat. Although some of the species of *Corynebacterium* are very slow growers, their importance on the skin is only now being recognized because of the

Human Microbiome Project. They are most commonly found near the apocrine sweat glands, of which the most notorious are in the armpit. Their food is the sweat from the gland, and they process it efficiently. But the byproducts of consuming the sweat are compounds that smell terrible to humans, making these bacteria the prime cause of body odor.

The highest diversity of bacteria resides on the dry areas of the skin, and is almost always a mixture of the four phyla we have mentioned (Actinobacteria, Proteobacteria, and Firmicutes). What are the driest parts of our body? Our hands and our buttocks are among the driest, which is where most of the bacteria from these four phyla reside. One of the more surprising results of the Human Microbiome Project is that members of the Proteobacteria, which were once thought to reside only in the gut, can be found in abundance on the skin. This makes the drier regions of the skin more variable than the gut or the oral cavity even within the same individual human examined.

There is also a temporal aspect to microbial diversity on our skin. That is, the microenvironment of various parts of the body changes from day to day and hour to hour. Or at least we hope they do, when we engage in activities such as taking a shower, washing hands, or changing clothes. Bathing is one way to alter the skin environment in a pretty drastic way. As mentioned previously, exercise will often change the moisture content of a part of the skin and result in elevated temperatures. And when we engage with the skin—such as to apply cosmetics or oils like sunscreens—we alter the microbiota considerably. Cosmetics companies have logically gotten in on the skin microbiome scene. One company suggests that knowing which species comprise the normal skin microbiome will help in the creation of better cosmetics by indicating which bacteria are best to use as product supplements.

Any site on the skin that is partially occluded fares better at maintaining a constant and consistent environment. Sites such as the ear canal and the nostril have been shown by the Human Microbiome Project to be very constant environments and hence to have highly predictable microbial communities. The sites of skin that change the most over time, such as the forearm, the bottom of the foot, the spaces between the fingers, and the back of the lower leg, by contrast, show more variation than even the oral cavity.

There are many microbes that can be pathogenic threats if they land on the skin. The Human Microbiome Project has divided these into three categories: disorders that are directly correlated to known microorganisms; disorders that are caused by a community of unidentified microorganisms; and disorders for which the causative organism is one that is typically commensal (it lives on us, and so gains from us, though does not harm us), but has gone bad. The first category includes well-studied disorders such as acne, eczema, and specific kinds of dermatitis. The second includes disorders such as infections from burns and chronic wounds. These are usually the result of colonization of the disturbed area by communities of bacteria. The third is perhaps the scariest and includes infections caused by skin bacteria that are usually benign hitchhikers on our skin. One of these, *Staphylococcus epidermidis,* happily lives a commensal life on our skin. It gains, but we don't lose. If the Dr. Jekyll–like *S. epidermidis* gets under the skin, however, then it can morph into Mr. Hyde, and some terrible infections can occur. This microbe is the most common factor in infections of heart valves and catheters in hospital operating rooms. Certain strains of this species can form biofilms, which are very efficient at warding off the human immune system. *Staphylococcus epidermidis* seems to be particularly good at fending off antimicrobial drugs, too. This would be bad enough if the drug resistance were just associated with *S. epidermidis.* But because this species is also efficient at transferring DNA from itself to other closely related microbes, it sometimes can transfer the resistance factors to species like *Staphylococcus aureus*. This bacterium is already very nasty and virulent, and adding new antimicrobial resistance can cause *S. aureus* to become incredibly dangerous.

Staphylococcus epidermis also is involved in regulating the host's innate immune response. It can stimulate the immune response to eliminate other species of bacteria, such as *Streptococcus* species and its congener *S. aureus*. This interaction of our immune system with skin microbes is not unique. Skin disorders, such as certain kinds of eczema, are a result of an interaction of skin microbes with the immune system. In this case though, instead of working with the immune system to eliminate other microbes, the microbes involved with eczema inhibit the production of antimicrobial peptides (AMPs). The lack of AMPs then down-regulates, or slows, the production of proteins involved in the innate immune response.

What Do Jammers and Keyboards Have in Common?

On February 10, 2012, in Eugene, Oregon, several women's roller derby teams squared off in an annual tournament known as the Big O. But this was no ordinary roller derby tournament. The players in the matches were being monitored for the microbes that they had brought to the tournament as well as for the microbes that were going to accompany them home after they had mixed it up with their rivals. The players from three teams (the Emerald City Roller Girls from Eugene; the Silicon Valley Roller Girls from San Jose, California; and the DC Rollergirls from Washington, D.C.) lined up, and scientists at the University of Oregon swabbed the microbes from their forearms.

Each bout, or match, in a roller derby is sixty minutes long and features two-minute segments called jams, during which the players skate around an oval flat track and try to score points by having a predesignated skater, called a jammer, skate past the opposing players. To prevent a jammer from scoring, the other teammates form a pack of "blockers" that usually consists of players from both teams jostling, roughing, elbowing, and pushing each other. It is strenuous exercise involving a lot of sweat and body contact, and usually all members of a team are in the mix at some time or another during the bout.

Samples were taken from players on each of the opposing teams both before and after two bouts. (The track floor was also swabbed to see what kinds of bacteria were present in the arena.) The first bout involved the team from Silicon Valley and the team from Emerald City. Four hours later, Emerald City played the DC Rollergirls. The University of Oregon researchers, James Meadow and his colleagues, had some very specific research questions they wanted to address:

1. Did team membership have anything to do with microbiomes?

2. Did microbiomes change during a bout?

3. If the original microbiomes of the three teams were different, did the microbiomes of the teams converge after the bout?

Why would roller derby competitors be of interest to scientists? It turns out that understanding how humans transmit microbiomes during everyday contact is a critical cog in our understanding of the interactions we have with our microbes. The roller derby offered an exaggerated version of how humans interact with each other and hence allowed a look at the phenomenon in real time. In other words, the researchers had found a good setting for studying how microbes might move about in the face of a lot of human contact. They had the baseline microbial makeup for two teams before they even made contact with each other, and they had an intermediate step from two teams who had made contact with other teams and were about to make contact with each other. And, finally, they had the same two teams after they made contact with each other.

After the University of Oregon research team had collected the samples, they returned to the lab and began the task of isolating DNA from the swabs that contained just some of the millions of bacteria living on the forearms of the roller derby girls. The 16S rDNA barcode gene was sequenced and analyzed with computational methods from the massive amount of DNA to give a list of the microbes living on the forearms of the contestants and a count of the relative amount of each kind of microbe there. Using a very visual approach, the researchers could map each roller derby competitor's skin microbiome onto a graph. The two axes represented measures of bacterial composition (which we will discuss shortly) such that if two of the girls' microbiomes were very similar in composition, they would end up in very similar spots on the graph. If they had very different microbiomes, then they would end up on very distant parts of the graph.

The results of the study demonstrated a recurrent theme about microbiomes of humans from different geographic localities. People from different areas of the United States and from different parts of the world have distinctive microbiomes that can be measured and compared. For instance, the DC Roller Girls were the only team that had a kind of bacteria called *Brevibacterium*, which is known to cause foot odor. The Silicon Valley Roller Girls carried two unique kinds of bacteria on their skin—*Alcanivorax* and *Xanthomonas*—while the Emerald City Roller Girls had three kinds of bacteria unique to them, *Dietzia*, *Coprococcus*, and *Alcaligenes*. Microbes can follow very strict geographic lines, and most likely the result was distinctive because the three teams were from very different geographic regions (especially the DC Roller Girls team). It makes sense that environmental variations, such as differences in the water, food, and soil, as well as temperature differences, would lead to these distinctive results. And because the girls from each team practice with each other, ride in the same cars to the match, and often hang out together (including to celebrate their victories or lament their defeats) their microbial inhabitants, and thus their microbiomes, become homogenized among the team members. That is, each team had a distinctive microbial profile, and the differences between teams were apparent. The graph clearly shows that the Emerald City team, the Silicon Valley team, and the DC Roller Girls each had very different starting microbiomes (Figure 3.2).

After the first bout, the teams' microbiomes shifted a bit. The one team that played in two matches (the home team, Emerald City Roller Girls) showed a shift in microbiome from the beginning of its first game to the beginning of its second game. In fact, after all of the pushing, shoving, elbowing, bumping, and falling, the microbiomes of all of the teams started to converge.

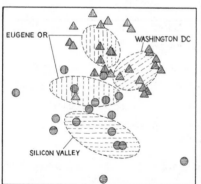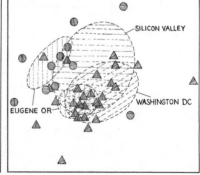

Figure 3.2. Multivariate plots of roller derby players' microbiomes, with Eugene, Oregon's represented by light triangles, Washington, D.C.'s shown as dark triangles, and Silicon Valley's marked with circles. The plot on the left shows the microbiomes before a "bout," and the one on the right shows the microbiomes after the bout. Note the tighter clustering of points after the bout.

The reason they converged is not so easy to tease apart. One reason the communities of bacteria might show convergence is that all of the players were exercising, so they all had an increase in body temperature and changes in moisture on the skin. This option can be rejected fairly easily, however, because the bouts were so short that exercise alone was likely not enough to implement the large-scale changes in bacterial composition seen by the researchers. A second possibility is that the arena environment (such as the microbes already in the air or on the floor) was strongly influencing the microbiomes of those people in the arena. In other words, the players picked up tons of new microbes as they skated around the arena. This possibility was the reason that the researchers collected swabs from the floor of the arena before the bouts. There can be substantial transfer of microbes from the environment of a building like an arena, because the players are continually being knocked to the floor and dust from the arena is stirred up not only by the players but also by spectators. The researchers, however, found that this possibility did not explain all of the shifts in microbes among the players. This leaves the most obvious reason—the constant contact of the players during the bouts. As the girls pushed and shoved against each other, microbes from their skin were transferred from one to another, until everyone had a similar community.

In another study that examined the potential of using skin microbiomes to identify individuals, researchers swabbed three keyboards used by three individuals in an office. Next they obtained the bacterial community profile of the keyboards and compared these with the bacterial community composition of the fingertips of individuals working in the office. The results of the comparison were stunning. Using basically the same approach as the roller derby study, the researchers were able to look at how the different samples taken clustered with respect to their similarity. Although there were some differences associated with the fingers used, there were very obvious differences between the individuals, implying that it would be possible to know who had used which keyboard—an astonishing feat.

Although the roller derby and keyboard studies might seem a bit wacky, they do help us understand some of the dynamics of how our skin microbiomes change and are influenced by contact with other humans. Our modern lives are quite different from those of our ancestors fifty thousand years ago. Cities have resulted in much closer human contact, and even instances of skin-to-skin contact, with complete strangers. The University of Oregon researchers suggest that "as the rise of mega-cities and population growth continues, humans may experience an increased rate of person to person contact mediated by urban living and global travel. To predict the implications of these changes will require, in part, understanding the ecological and evolutionary forces that act on the skin microbiome." By participating in the University of Oregon study, the roller derby girls have contributed to this understanding.

Belly Buttons

The umbilicus has more than likely been an object of admiration for ages. Perhaps the most famous belly button is in Leonardo da Vinci's Vitruvian Man drawing, where the belly button is the center of the radially reaching man. Today, creationists argue over whether Adam and Eve had belly buttons, since according to a literal reading of the Bible, neither Adam nor Eve had a mother. At least we all know where belly buttons come from. The umbilicus starts to form shortly after birth when the umbilical cord is snipped to finish the separation of the newborn from its mother. To cut it without making a bloody mess, the cord is clamped or tied off both close to the baby's stomach and about two inches farther away from the baby. The region between the two clamps can then be cut to sever the umbilical cord. The part of the umbilical cord still attached to the baby forms scar tissue that in about 90 percent of the population invaginates in such a way that the baby has an "innie," and in about 10 percent of the population scars in a different way and forms an "outie." Contrary to most beliefs, an outie is not caused by where the umbilical cord is cut, but rather is a matter of how the umbilical cord was originally attached to the baby during fetal development. In fact, there are many ways the umbilical cord can be attached during fetal development, and hence there are many different ways a belly button can be an innie or an outie. Because of the way the umbilical cord is attached and the way the cut cord heals, there are different shapes that a belly button can take, such as being T-shaped, or oval, or more like a vertical slit. In other words, the actual topology of the belly button can vary from person to person, not just with respect to innieness and outieness, but also with respect to size, depth, and shape.

Sal Mineo, a famous actor in the 1950s, apparently had a belly button fetish. This is not so far-fetched or perverted as one might think. Some researchers have studied the shape and appearance of belly buttons and have discovered that some shapes are preferred over others by humans. Aki Sinkkonen, a Finnish researcher, has taken a close look at how males react to female belly buttons. Sinkkonen has shown that outies are almost always considered unattractive. Also, really deep innies are considered less attractive than shallow innies. His explanation is that somehow men are using the shape of the belly button to judge how fertile a female is. Other researchers have suggested that the position of the belly button can be an indicator of athletic prowess. This is because, as Leonardo implied in his famous drawing, the belly button closely approximates the center of gravity of a human body. The higher the position of the belly button, the higher the center of gravity, giving a runner an edge. Since from a biophysical standpoint running is simply the act of falling forward, a higher center of gravity allows this process to be slightly more efficient and easily controlled.

Perhaps the most spectacular belly button research, however, is on the microbial biodiversity inside them. In 2012, Rob Dunn and his colleagues at North Carolina State University examined the belly buttons of science writers who had attended scientific meetings in the area and belly buttons of people who attended a scientific conference a month later. Their paper, appropriately entitled "A Jungle in There: Bacteria in Belly Buttons Are Highly Diverse, But Predictable," describes the results of using next-generation sequencing approaches to determine the number and kinds of bacteria living in the belly buttons of the sixty people in the study from the two different events.

The diversity that the researchers observed was spectacular—they found a total of nearly 2,400 phylotypes of bacteria in the sixty different belly buttons. (Remember that a phylotype is shorthand in microbiome studies for "species.") The grand majority of phylotypes (2,188) were found in fewer than 10 percent of the belly buttons studied. Even though there were two hundred or so phylotypes found in more than 10 percent of the belly buttons, none of these were found in all belly buttons, and only eight phylotypes were found in more than 70 percent of the belly buttons. The authors of the work suggest that the belly button biodiversity is on par with the animal diversity in a jungle, which brings us to a quote from the "queen of pop," Madonna. When discussing her belly button she once said, "If one hundred belly buttons were lined up against a wall, I could definitely pick out which one is mine." Although we don't doubt her ability to pick her belly button out of a lineup, an equally valid way to tell which belly button is hers would be to sequence her belly button microbiome.

Although these results might sound chaotic, they are, on the contrary, rather predictable. By asking whether looking at the first cohort of belly buttons (from thirty-five science reporters) can predict the composition of the twenty-five belly buttons from the second cohort, the researchers showed that there is a nice correlation. As the authors of the paper state, "Frequent phylotypes tend to be predictably frequent and infrequent phylotypes predictably infrequent." In addition, there were no big surprises in what resided in the belly buttons of these sixty people. Bacteria that tend to hang out on skin were the abundant inhabitants of the navel, and phylotypes from three major groups of skin bacteria—Corynabacteria, Staphylococci, and Actinobacteria—were found most frequently in the belly buttons. One surprise did come up with a bully button of a participant who claimed not to have showered or bathed for a long time. This person's belly button had two rather bizarre phylotypes from that group of microbes we discussed in Chapter 1, the Archaea. Remember that Archaea are called extremophiles, because they live happily in extreme environments. Apparently this person's belly button provided an extreme environment for these phylotypes of Archaea.

Perhaps the most significant result of the study is that there are an amazing number of rarer phylotypes living in the belly button. In addition, some people's belly buttons actually have three times the number of phylotypes than others. These two facts, in combination with research on the immune system, lead us to an interesting idea about how our immune systems might react to skin bacteria. Studies done to examine the influence of the skin microbiome on the immune system can lead to ideas about possible therapies to ensure more healthy development.

Acquiring a First Microbiome

Where did you get your skin microbiome? The answer is actually quite simple because of the way birth works in mammals and the way that bacteria colonize other organisms. The skin microbiome starts from scratch when a baby is born. The amniotic sac in which the fetus develops is a relatively sterile place. Once the baby enters the birth canal or is removed from the mother by Cesarean section, however, bacteria, viruses, and possibly fungi start to stick to the body.

The skin microbiome of the baby differs by type of delivery. In a study of several women giving birth in a Venezuelan hospital, researchers were able to compare vaginally delivered neonatal microbiomes with C-section neonatal microbiomes. The researchers took swabs of the mother's skin, vagina, and oral passage one hour before delivery. The oral passages, skin, and nasal passages of babies were also swabbed five minutes after delivery and again twenty-four hours after delivery. The mothers had very different microbial communities depending on the part of the body sampled, but the newborn babies had the same bacterial communities in all three of the places where swabs were taken. This result means that the microbiome of newborns is the same across the entire body, harboring the same phylotypes regardless of body part. But if a fetus is relatively sterile upon delivery, where does this initial microbiome come from?

The answer to the question is pretty straightforward. The neonate's skin and oral and nasal cavities simply attract nearly all the bacteria with which he or she first comes in contact. So for babies delivered vaginally, the first microbiome they encounter is the mother's vaginal microbiome, and indeed babies delivered in this way have an undifferentiated microbiome in all parts of their bodies that is extremely similar to their mothers' vaginas. By contrast, if a baby is delivered via C-section, it bypasses the vagina of its mother. Its first contact is with the air of the outer world and the skin microbiomes of people in the delivery room. Hence the newborn delivered by C-section has an undifferentiated microbiome that looks like a mixture of skin microbiomes and organisms floating around in the delivery room. The study was not precise enough to determine where exactly the C-sectioned neonate's microbiome originated—just that it comes from both maternal and nonmaternal surfaces but looks more like a skin microbiome than anything else.

What is also clear is that the newborns delivered by C-section are more susceptible to infections. For instance, of all the babies who develop *Staphylococcus* (or "staph") infections after delivery, as many as 80 percent are from C-section deliveries. This observation, coupled with the results just described for microbiomes, might mean that the community of microbes that newborns obtain from their mothers' vaginas might have some protective function with respect to staph infections. The vaginal microbiome is packed with phylotypes from a specific kind of bacteria called *Lactobacillus,* which don't allow other microbes, including *Staphylococcus,* to establish themselves and hence cause infections. The more skinlike microbiomes of neonates delivered via C-section, by contrast, allow for the establishment of *Staphylococcus* and for staph infections to occur.

Finally, the initial seeding of a baby's skin and oral and nasal microbiome with specific kinds of bacteria will influence the way that the rest of the body obtains the differentiated microbiomes with which we eventually end up. Such succession is very important for establishing a working gut microbiota, for proper interaction with the immune system, and for adequate early nutrition. Right now we have comparisons only of vaginally delivered versus C-section delivered neonatal microbiomes. More work on this subject will be useful in ensuring that newborns get the right chances to develop working gut microbiomes and immune systems.

Studies on how we got our first microbiome tell us a lot about how microbiomes are seeded and how they first develop. But how do our skin microbiomes change over time? To understand this question, we need to ask what is out there around us.

Subways

Ana Gasteyer, a former *Saturday Night Live* cast member, probably rode the subway to the 47th to 50th Street stop on the Sixth Avenue line in New York City on the way to her job at 30 Rockefeller Center. Of the subway she said, "Even if you're alone on the subway, you never feel lonely." She couldn't have said it better, because the New York City subway, both psychologically and, as we will see in a moment, physically, always has its share of other passengers. Norman Pace, the biologist mentioned in Chapter 2 as one of the founders of microbiome studies, actually started his work in this area because he was interested in the kinds of microbes that live in different environments. The approaches for looking at microbiomes were first honed on environmental samples, like those from previously uncharacterized environments such as hot springs and the Sargasso Sea. So it has not been especially difficult to make the transition to looking at microenvironments like houses; places where lots of different people come into contact with each other, such as a subway or a museum (or roller derby); or the human body. Even objects that we humans use on a day-to-day basis, such as our shoes or our cell phones, might tell us something about the microbes with which we come into frequent contact.

To this end, microbiologists have investigated the microbial diversity of several subway systems. New York City has 468 subway stations and a ridership of over 1.6 billion a year, and is being examined not only for air quality (which might affect the microbial communities in us) but also to take a microbial census from different parts of the subway system that humans touch daily. Norman Pace has used specialized air collectors called fluid impingement devices, which were designed to be inconspicuous so that average New York City subway riders—who have been highly conditioned to "say something" if they "see (or hear) something"—would not take notice of the survey. Pace's team sampled the subway system in three stages over a one-and-a-half-year period. They sampled platforms along the 4, 5, and 6 lines (the green line in the New York City subway system), including several stops downtown from 42nd Street. They also used two nonsubway controls—upstairs and outside the Union Square station and in Grand Central Terminal on the mezzanine in the great hall. They also sampled one unused platform during the duration of the study.

Was the microbial makeup of the New York City subway system air generally more diverse than that of a typical belly button? The results of the Pace study revealed a paltry diversity of bacteria in the subway system, and in fact the majority of microbes were fungi. More surprisingly, the microbiota of the subway platforms differed very little from the outside environments. Why? Because of the modernized air exchange systems of the New York City subway system. Most notably, however, was the observation that pathogens were absent over the course of the survey. About 5 percent of all of the bacteria found in the subway were typical human skin bacteria, which means that humans are shedding the bacteria from their skins in a pretty regular way and might contribute to the eventual homogenization of the skin microbiomes of New York subway riders. By comparing the kinds of bacteria at the unused station to the used stations, they found that these stations also had the skin bacteria, which indicates the subway system itself is becoming a distinct environment as a result of forced airflow throughout the system.

Other subway systems have been examined in detail using next-generation sequencing approaches, as well as "culture-dependent" approaches, where the air of the subway system is analyzed by putting it onto agar plates. Subway systems in Seoul, Korea; Tokyo, Japan; and Oslo, Norway, have also been analyzed, and their microbiomes indicate a stability of microbial inhabitants similar to that of the New York City subway system microbiome.

In a move away from analyzing bacteria in the air, Weill Cornell Medical College researcher Chris Mason also studied the microbes that live on objects in the New York City subway stations. Using undergraduate-student "swab squads," this project, called the PathoMap Project, took samples from all of the 468 stations in the system. Each station was swabbed in many places—at the ticket kiosk, on a turnstile bar, on the surfaces of benches and trashcans, and at several locations on a train stopped at the station (Figure 3.3).

The swab on the subway train is most often taken from a metal surface such as a rail that passengers hold to stabilize themselves during the bumpy subway rides they experience every day. The re-

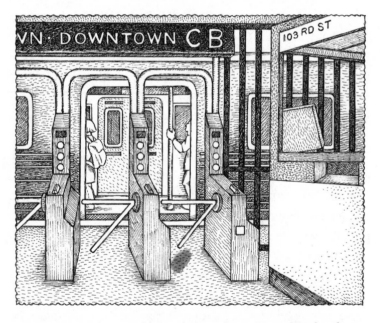

Figure 3.3. A typical New York City subway station entrance. Samples of the microbial communities in this system are being taken from the ticket kiosk, turnstiles, and benches on the platform and from the hand rails in the subway car.

sults of the study seem to indicate a somewhat different microbiota than the aerosols, with skin bacteria, specifically species in the genus *Acinetobacter*, found in abundance and some *Streptococcus* bacteria found less abundantly so. There is also a larger than usual frequency of *Enterococcus* bacteria present on surfaces. This discovery is slightly disturbing, because these bacteria are common inhabitants of feces, and the presence of this microbe on subway surfaces more than likely indicates poor hand-washing habits by riders.

The PathoMap Project is interesting not only because of the information it will yield regarding the diversity of microbes in a heavily used human habitat, but also because of the way it is being accomplished. Completing the project will probably require over a million U.S. dollars, so the researchers conducting the study have opted to fund part of it with an online crowd-funding approach.

Although the 1.6 billion people who ride the New York subway annually provide an amazing sample size for understanding the microbial diversity of a heavily used public space, perhaps examining the patterns found in a more stable public space such as a museum will reveal information about which microbes people encounter in this type of environment. For example, researchers have examined the microbiome of the Louvre Museum over time to determine the stability of microbial communities. For a six-month period, French researchers sampled Room 36 in the Richelieu wing, where the work of the French Regency master Salle Watteau hangs.

Using culturing and next-generation sequencing techniques, the researchers discovered that the amount of *Escherichia coli* and the fungus *Aspergillus fumigatus* was relatively constant during the six months. In addition, three of the samples (one from day 1, one from day 157, and one from day 164) were analyzed using next-generation sequencing approaches (Figure 3.4). Several major families of bacteria were found for each day, but the microbial composition of the

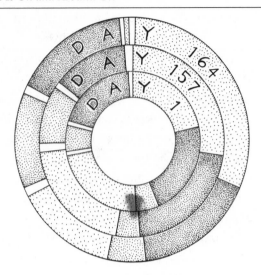

Figure 3.4. Circle diagrams showing the stability of microbial communities in Room 36 at the Louvre Museum in Paris. The samples were taken over a period of 164 days. The numbers on the circles indicate which day in the sampling period the sample was taken.

samples was remarkably similar across the three days. These results indicate that the Louvre Museum, or at least Room 36 of the Louvre, has a very stable microbiome.

What about other rooms in public spaces? Although we can testify that the bathrooms in the Louvre Museum are pretty clean, we are sure our readers have encountered a restroom or two that seemed too filthy to use. How microbes establish themselves and coexist in public restrooms has been studied by the same researchers at the University of Colorado who developed the handedness study described at the beginning of this chapter. By looking at the microbiomes of male and female restrooms on the Boulder campus, the Colorado researchers were able to get an interesting first look at which microbes are in public restrooms and how they might be interacting with humans.

The researchers swabbed ten surfaces in these restrooms (door handles into and out of the restroom, handles into and out of a restroom stall, faucet handles, soap dispenser, toilet seat, toilet flush handle, floor around the toilet, and floor around the sink). This sampling protocol resulted in sixty men's room samples and sixty women's room samples, all of which were then examined using next-generation sequencing approaches. These college bathrooms were pretty diverse with respect to the different kinds of bacteria present. The four most prevalent phyla of bacteria found in the bathrooms were Actinobacteria, Firmicutes, and Proteobacteria. If these sound familiar, they should be, because they are the same four phyla that predominate the habitats of our skin. The researchers found that samples taken from a floor were very similar to the other floor samples no matter where the sample was taken. Likewise, the toilet samples had a similar profile. And finally, samples of places that hands touch (other than parts of a toilet) formed a cluster. Of these three major clusters, toilets, specifically toilet seats, had loads of microbes commonly found in fecal matter, suggesting a lot of contamination from feces in the restroom to the seat. Floor samples were populated by large amounts of bacteria commonly found in soil samples, indicating that shoes were a big factor in populating the restroom microbiome. Finally, wherever hands were involved, the microbes most commonly found in the skin microbiome were found in large quantities.

The major difference between the male restrooms and female restrooms was the occurrence of microbes from the genus *Lactobacillus* in the female restrooms. This result is not surprising, because species of bacteria in this genus are associated with the female reproductive tract. Overall, the composition of the microbiome in restrooms was diverse, but directly influenced by the parts of the human body that come into contact with the surfaces of the restroom. And because restrooms serve as a sink, so to speak, for microbes from outside sources such as shoes, hands, and skin, they are an important and immediate subject for studying human hygiene and perhaps even for modeling human habits involved in hygiene.

Home Is Where the Heart, and the Microbes, Are

Although people who live in cities with subway systems might find these earlier results interesting, those of you who do not encounter subways routinely still need to think about what is in the air and the buildings around you. One of the more important places to be characterized with respect to microbiomes is one's home. And once again it is researchers in North Carolina who have done the initial approach to characterizing the "houseome."

If you were trying to find microbes in your home, where would you look? Making sense of the kinds of bacteria that live in your home might appear daunting at first. To us, our homes are pretty diverse places (unless you live in a four-hundred-square-foot studio in a city). To microbes, the diversity of places to live in a home must be enormous. So the researchers in this study had specific hypotheses they wanted to test. For instance, they were interested in whether regular cleaning had an impact on the diversity of microbes in the home. They took the approach of treating the house as a collection of microhabitats and examined surfaces that were cleaned regularly and those that had evaded cleaning for long periods of time. Temperature and humidity are important habitat parameters for microbes, too, so they decided to look at habitats in the home with these differences as well. Who and what live in the house are also important. We have learned that subways, museums, and public restrooms all are influenced by the skin microbiota of the people who frequent them. The organisms that are roaming around the house are important parameters of habitats in the home, too. The various places tested in the forty homes are shown in Figure 3.5.

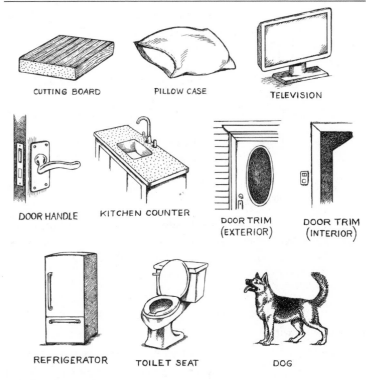

CUTTING BOARD   PILLOW CASE   TELEVISION

DOOR HANDLE   KITCHEN COUNTER   DOOR TRIM (EXTERIOR)   DOOR TRIM (INTERIOR)

REFRIGERATOR   TOILET SEAT   DOG

Figure 3.5. Sampling sites for the houseome study.

Although the dominant phyla in the home appear to be relatively consistent and mostly from the Proteobacteria, Firmicutes, and Actinobacteria groups, the profiles of specific habitats in the home are pretty interesting (Figure 3.6). On average, more than one hundred different phylotypes were found in each of the nine different habitats examined, which means that each habitat has many phylotypes associated with it. It turns out that surfaces that are regularly cleaned such as a kitchen cutting board and—believe it or not—a toilet seat have fewer phylotypes than areas that are not cleaned regularly such as a television screen, or the trim around the door of a house. In

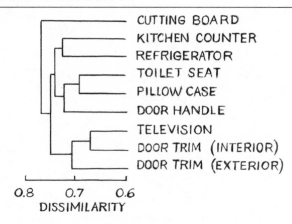

Figure 3.6. Network showing the similarity of the various sampling sites in the house microbial community study. Note that the toilet seat and the pillow case are more similar to each other than they are to any other surface.

addition, surfaces that are cleaned regularly seem to be more similar to each other with respect to the kinds of microbes present.

This result is consistent with another more detailed study of kitchens that examined the hypothesis that cleaning is an important arbiter of diversity in kitchens. The researchers in this study considered the kitchen to be a set of ecological habitats, and so examined in great detail the movement or succession of bacteria around the room. They found that places such as exhaust fans in kitchens harbored greater microbial diversity than places that were cleaned regularly like sinks. Although these observations make sense, what they add is a quantitative approach to understanding the movement of microbes in one of the more important rooms of our homes. For instance, the succession of bacteria from the skin to recently cleaned surfaces can be inferred. In addition, rare microbes can also be followed. In the more detailed kitchen study, microbes from genera that harbor food-borne pathogenic bacteria were found throughout the kitchen—but in very specific distribution patterns. These patterns could potentially be examined as part of efforts to promote better kitchen hygiene.

The houseome study also yielded some results that seem intriguing at first. One that we find rather disconcerting is that when the phylotype composition of the different surfaces is examined for how similar the nine surfaces are to each other, a pillow case is most similar to a toilet seat (see Figure 3.6). But although this result might seem a bit gross, both a toilet seat and a pillow are in direct contact with large areas of human skin. Whereas the microbiome of the buttocks is a bit different from the microbiome of the face, both harbor the same general kinds of skin bacteria, and hence these two surfaces have similar phylotypes stuck to them.

Why different homes vary in the distribution of microbes also has a lot to do with what comes in and out of the house. As has been demonstrated, a lot of the skin microbiome gets shed into the home. In addition, microbes can travel into the home via articles of clothing such as shoes and items we carry around with us, like wallets and cell phones. The Home Microbiome Study (homemicrobiome.com), funded in part by the Alfred P. Sloan Foundation, has expanded research in this area by collecting information from many, many "citizen scientists"—ordinary people who are willing to help—about what is in their homes, on their cell phones, and on their shoes. The results demonstrate that, in general, the microbiomes of phones have less diversity than the microbiomes of shoes, which makes sense since cell phones are in contact with our hands and places where we lay them down, whereas shoes are in constant contact with wherever we walk. Another project within the same study uses the citizen science strategy to characterize a large number of homes for microbial diversity throughout a living space.

One factor that the researchers took into consideration was the presence of pets in the home. They found that homes with dogs had different microbiomes than homes without. This result was in line with another study done two years earlier, where the presence or absence of man's best friend influenced the diversity of house dust.

Animals have more than just fleas and ticks on them. In fact, a dog's skin microbiome is every bit as complex and maybe even more so as ours, and every bit as responsive to outside influences as ours. A 2014 study by veterinarians and microbiologists revealed that a dog's skin is a complex collection of ecological habitats for microbes, just as ours is. By sampling from several regions of a dog's body, researchers at Texas A&M University in Texas were able to characterize some of the basic parameters involved in bacterial diversity associated with the external surface of a dog. Sites on the dog's body where mucosa exist—for example, the nose—have different microbiomes than sites that are hairy. And both of these kinds of sites are very different from the perianal region. Hairy regions are by far more diverse than these other two general areas. The researchers who did this study examined several healthy dogs and several dogs with skin disorders (like allergies or fleas) and showed that skin regions of healthy dogs are more diverse with respect to microbes. Another result that kind of mirrors what researchers see in humans is that there is considerable variation from individual to individual in the microbiome makeup. This means that individual dogs might be identifiable from their skin microbiomes, just as researchers have claimed is true for individual humans.

Because dogs, and pets in general, have unique and copious skin bacteria and skin is a good way to seed microbial communities, then it might be that pets could influence the microbiome of the home. The North Carolina researchers did indeed examine this possibility and found that homes with dogs had different microbiomes than those without dogs (even those without dogs but with cats). Specifically, households with dogs had microbiomes that looked more like dog skin microbiomes. After considering all we have reported about how skin bacteria can influence the microbiome of the house, this should not be surprising. But the results of this study and another on the influence of dogs and cats on air microbiomes in houses indicate that having a pet makes a world of a difference in how we humans respond to the air in our houses. In particular, being exposed in the home to the microbiome of a pet may help keep members of the household from developing allergies.

Lest we think that getting out of the house and to the office gets us away from the multitude of microbes out there, recall the museum study at the Louvre—or other studies that examined workplace microbiomes. Krissi Hewitt and colleagues at the University of California, San Diego and the University of Arizona examined the microbiomes of offices in three metropolitan areas—Tucson, New York, and San Francisco. Hewitt and colleagues swabbed five areas in thirty different offices in each city—chairs, phones, computer mice, computer keyboards, and desktops—and noted whether the office inhabitant was male or female (Figure 3.7).

Although it involved a small and rather limited sample, the study did demonstrate some interesting trends in how microbiomes are formed and maintained. There were two ways the researchers in this study characterized the microbiomes. The first involved simply counting the number of microbes on surfaces and using the variables of city, surface, and gender of office occupant to further describe the many microbes on the different surfaces. The second method ignored the level of microbial presence and, much like the houseome studies, determined how diverse the microbiomes were.

The results? Phones and chairs had more bacteria than the other three surfaces sampled in the office. Second, the surfaces of men's offices were teeming with more microbes, but whether this was simply because men in general have larger surface areas or because they are bigger slobs than women was uncertain. The diversity estimates for this study indicated that, once again, skin bacteria from inhabitants

Figure 3.7. Typical desktop showing the places where samples were taken in the office microbial community study. Chairs, phones, computer mice, computer keyboards, and desktops were all sampled.

of the office were the major determinant of the microbiome diversity of the offices in the three cities. There were also some regional differences in the pattern of diversity. San Francisco had lower levels of microbes than either Tucson or New York City, but Tucson had a distinguishable microbial community, whereas the microbial profiles from the offices in the two cities with sports teams named the "Giants" were very similar. This pattern can be explained by the desert soils around Tucson being quite different from the temperate, highly urbanized environments of both San Francisco and New York: when office workers arrived with soil on their shoes, different microbial communities transferred from their soles to the office itself.

This all begs the question of how the microbes make it into rooms like offices, museum galleries, and classrooms. Do they sneak in with the wind? How do they travel? A study of classroom microbiomes at the University of Oregon in 2014 is enlightening in this regard. The researchers in this study wanted to determine how the movement of a lot of humans in and out of a heavily used and dynamic space affected the dispersal of microbes within the classroom. They reasoned that the bacteria residing on surfaces could be spreading throughout the room either by migrating across adjacent surfaces or by hitching rides on the students who were moving about the class. By examining four distinct types of surfaces in the classroom, the researchers were able to determine that the human factor was much more important than the proximity of surfaces. In fact, humans can transmit surface bacteria even without direct contact to a surface. Every moment of the day and night, we are shedding our skin microbiome like crazy.

CHAPTER 4

What Is Inside Us?

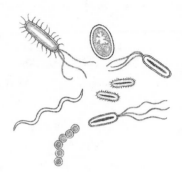

Frightening body invasions are a mainstay of popular science fiction movies. Who could ever forget their first time watching the movie *Alien* and seeing the space creature emerge from the crew member's chest? The discovery that humans are merely a habitat for the larval stage of this alien is both surprising and gross. But perhaps the most horrifying invasions of our bodies in sci-fi are those caused by infections that induce zombies. With zombies, human bodies are usually invaded by a viruslike agent that instead of causing death turns the human brain to putty while the body lives on to crave and hunt human flesh. Even humans who are not yet full-fledged zombies become moving, fleshy Petri dishes for the growth of the pervasive microbe until they die and become zombies themselves. Perhaps one of the most terrifying scenes in the television series *The Walking Dead* is not when zombies are chasing and eating living humans, but rather when the hero learns that all humans are infected with the agent that produces "walkers." For many, this form of body snatching by some outer force might seem a little too real for comfort. In fact, the fear that our bodies will be invaded has led to the so-called war on microbes that has existed since Koch and Pasteur's time. For the past century or so, we have focused on the elimination of microbes from our bodies as a strategy for treating infections and many other maladies. But we are lousy with microbes that live inside us, and we are living in a great degree of harmony with those microbes, much in the same way we live in harmony with those microbes living on us. Our normal interactions with both groups are more like "détente" than war.

Microbes within us vary significantly from those upon us. The microbes on us interact with only a single major organ, albeit the largest one in our bodies—the skin—and the mechanisms that our bodies and the bodies of other mammals have evolved to deal with them are mostly mechanical and armorlike. The habitats that exist on our bodies are diverse and provide niches for a large number of microbial species, but these dozen or so skin cell types are just variations on a theme.

By contrast, the cell types in our bodies differ radically, ranging all the way from enucleated red blood cells to the highly differentiated cells of our nervous system. For instance, the human body has dozens of internal organs, and hence hundreds of different cell types for these internal organs. Animals such as Placozoa (a basal form of invertebrate) and sponges (also called lower Metazoa) have only four and eight distinct cell types, respectively. Cnidarians, another lower metazoan, are not much more complex, with just over a dozen cell types. Insects are a step up, with more than twenty-five cell types. Because there are many more cell types in the human body, there are also more niches for any interloper to explore and colonize. There is also a wide variety of cell systems, such as the nervous system and the circulatory system, which can allow for the movement of microbes into and throughout the body. All these factors together make the interior of the human body a playground for microbes.

Microbes enter the human body in four ways. First, entry can be gained through ingestion of food and liquids, where the mouth is the gateway, and the esophagus, stomach, and intestines provide the major colonization sites. Second, microbes gain entry into the body through the respiratory system when we breathe. Third, a microbe can gain entry into a body through the blood, usually through cuts, transfusions, intravenous drug use or abuse, or insect bites. Finally, entry can be gained through the genital and anal openings through sex.

The Oral Microbiome

A microbe or group of microbes attempting to colonize through the oral route faces a wide variety of environmental parameters. Two reasons that the gut microbiota are so variable are pH (which measures acidity) and the sheer numbers of microbial interlopers. Figure 4.1 shows the wide range of environmental conditions in the digestive tract past the esophagus. The oral cavity is one of the areas of the body populated by the most amazingly diverse array of microbes, with microbes from the two main microbial domains of life residing there—Archaea and Bacteria—and, of course, viruses. Eukaryotes such as fungi and amoeba are also living in there sometimes. Nasty fungi such as the yeast *Candida,* which causes the throat infection called thrush, are among some of the eukaryotic residents. For the most part, though, the oral microbiomes of people are rather healthy assemblages of cohabitating microbes.

The first step in characterizing any microbiome is to see what species are living there, and the oral microbiome's species composition has been described in detail thanks to the Human Oral Microbiome Database (HOMD). Nearly 1,200 species of bacteria have been categorized so far, and, of these, over half are what microbiologists call prevalent; in other words, they are not rare. In fact, these 1,200 oral species are distributed among thirteen phyla of bacteria, making the oral microbiome about three times more variable

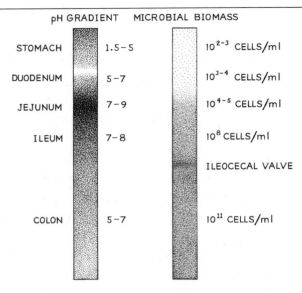

Figure 4.1. The huge variations in pH and microbial biomass in the human digestive tract.

at the phylum level than the skin microbiome. But although most of the microbes that have been named and cultured are associated with humans, only 24 percent of the microbes in the oral microbiome are sufficiently well known to have been named by bacterial taxonomists. An additional 8 percent have been cultured but not named, leaving about 68 percent that have no names and have not been grown in culture. This lack of knowledge about the species of microbes in our mouths is striking. By comparison, an entomologist going into a forest in the eastern United States would know the names of nearly 99 percent of the species of insects she encountered, and would be shocked to see a new species. And here is the human mouth, where we have no taxonomy for almost three quarters of the organisms living there.

Because the oral cavity is what an ecologist would call an open system, with materials or organisms coming in and going out all the time as a result of breathing and ingestion, it is rather difficult to determine which bacteria are living in the typical oral cavity. Researchers have relied, then, on the idea of a set of bacteria that appear to be present in as broad an array of individuals as possible. They call this a core microbiome, and the assumption is that this is the minimal set of microbes that comprise a healthy oral microbiome. Next-generation sequencing approaches on several sampled individuals have indicated that in the typical oral microbiome there are, on average, about 250 species. In the entire sample, there was a total of about five hundred species. There is approximately 70 percent overlap when any two individuals are compared, but when all the samples are compared together, the overlap drops to 33 percent or so, meaning that about 150 species make up the core microbiome.

To see if the environmental factors that people experience during their lives affect the mouth microbiome, researchers periodically sampled the oral cavities of individuals over a ten-year period (between ages twelve and thirteen, again between the ages of seventeen and eighteen, and finally between ages twenty-two and twenty-four). The sampling group included twins—both identical (monozygotic) and fraternal (nonidentical or dizygotic)—as well as a cohort of nontwins. The results indicated that twins, both identical and fraternal, have more closely related oral microbiomes when they are young than the cohort of nontwin individuals. If there is a strong genetic or hereditary component to the composition of the oral microbiome or any microbiome for that matter, then identical twins should have more similar oral microbiomes than fraternal twins. Such a difference was, however, not observed. In addition, as the twins grew older and their lives (and perhaps their eating and oral hygiene behaviors) became more distinct, their oral microbiomes diverged.

There is one human activity that directly affects the oral microbiome—kissing. The first record of kissing is 3,500 years old and written in Sanskrit, but kissing is certainly a much older activity than that. In fact, one of the explanations that anthropologists use to describe the origin of kissing is called "kiss feeding"—a behavior in primates (and other animals) that was and still is quite prevalent in many human cultures. Kiss feeding is the pre-mastication of food by a parent to soften the food for an infant, and involves transferring the softened food from the parent's mouth to the child's. Another hypothesis that anthropologists use is that kissing is intuitive. Whether one hypothesis or the other is right, kissing is an efficient way to transfer microbiomes between kissers. Studies have shown that couples living together have very similar oral microbiomes. Whether this is because of kissing or because of the couples living in similar household environments was addressed by a Dutch study that followed several couples engaging in intimate, "French" kissing. The Dutch researchers even performed experiments on the kisses and kissers. They found that an intimate kiss did not significantly increase the similarity of kissers' oral cavity with respect to their microbiomes. What they did find, though, is that the salivary microbiomes became more and more similar between two kissers, giving the saying "swap spit" a literal meaning. In addition, they showed that some parts of the mouth would change rapidly after kissing had ceased and others, like the upper surface of the tongue, were very stable. Using a controlled experiment with yogurt, which the researchers loaded with *Lactobacillus* and *Bifidobacterium*, they also determined that on average more than eighty million bacteria are transferred with each ten-second French kiss. All of this suggests that if you are married and thinking of engaging in an affair, you will have to make sure that no intimate kissing is involved, or that you do not allow your oral microbiome to be sampled after the dirty deed, for it will give you away.

The geographic and phylogenetic components of the oral microbiome have also been examined in detail. In a broad geographic study of 120 people from twelve different parts of the world, researchers discovered that, although individual microbiomes varied even within geographic localities, there was no clear pattern of geographic differentiation, other than that the farther from the equator they were, the more similar their microbiomes were. (This study did not attempt to take into account cultural background or ethnicity.) Contrast this result with the ability of the houseome researchers to differentiate between the microbiomes of toilet seats and doorways.

For comparison, consider the microbiome of our closest living relatives, chimpanzees and bonobos. Bonobos are also known as pygmy chimps and have only recently been recognized as a distinct species from the common chimp. Our lineage diverged from these two species almost 7 million years ago, whereas bonobos and common chimps diverged from each other about 2.5 million years ago. By contrast, all humans on this planet can be traced backed to common ancestors who lived only about 180,000 years ago.

In a study, researchers compared a set of chimpanzees that lived in an animal sanctuary in Africa with some of the humans who worked there. They also examined a set of bonobos and humans working at a different African sanctuary in another country. In addition, they examined a number of chimpanzees from the Leipzig Zoo in Germany, which had been in captivity for a long period. The results revealed a strong species-specific component to the oral microbiome—in other words, the chimp and bonobo microbiomes were more similar to each other than each was to the humans with whom they interacted. In addition, the zoo chimps had a higher degree of variability in their oral microbiomes compared with chimps and bonobos living in relatively wild conditions in Africa.

These results suggest that we need to look at our association with our oral microbiome through two lenses. The first lens is a phylogenetic or genealogical one: there is a clear association of microbes with us, because they have coevolved with us. But the second lens is the "nurture" or environmental lens, which includes the influence of habitat. The first lens leads to the identification of a core of microbes that more than likely have coevolved with humans and have established mutualistic relationships with us. And the second lens allows consideration of the variation around the core, and offers the opportunity to use microbiomes to identify individuals in populations.

Tonsils and Teeth

Farther back in the oral cavity are the tonsils and the adenoid glands. It shouldn't be too surprising to learn that the healthy microbiome of the tonsils is very similar to the rest of the mouth and the throat. In fact, in a healthy person the tonsils, dorsal tongue, and throat have nearly identical microbial species compositions.

One study of a child with chronic tonsillitis may help illuminate the potential origins of infection in this area. When the child's tonsils, adenoid gland, and inner ear passage were typed for their microbiome, researchers found that all three sites were not as variable as other regions of the body, but each did have its own unique disease state microbiome. The inner ear and tonsils shared the least number of species of bacteria, while the adenoid seemed to sit right between the other two sites. The study suggests that the adenoid may be the source for the microbiota that eventually end up in the tonsils and the inner ear, two places that most parents of small children know are frequently sites of infection. Removal of the adenoids (called an adenoidectomy) might sound like a good idea to get rid of the pathogenic microbes. But not so fast. These glands also probably modulate the normal cadre of microbes in the inner ear and tonsils, so their removal might cause ecological disruptions elsewhere.

Pig tonsils are a popular site for microbes, because this tissue in pigs is rather large compared with that in humans, the tonsils are in the path of ingested food and air for the pig, and infections in this tissue are chronic in pig populations. Their tonsils are a sponge for any microbe entering the porcine body, and they harbor a very different microbiome from that found on human tonsils. The major microbial residents on pig tonsils are bacteria from the family Pasteurellaceae, while humans harbor relatively very few of this family of microbes on the tonsils. Humans instead harbor species of *Streptococcus* and *Staphylococcus* as well as a number of species from the Firmicutes. These differences between pig tonsils and human tonsils are no doubt a result of our phylogenetic divergence from each other and the coevolution of microbes with this tissue.

We have left one region of the mouth until last—our teeth. The evolution of teeth in mammals is a fascinating subject. Some mammals, like anteaters, have completely lost their teeth, whereas others have teeth but have lost the enamel that coats and strengthens the outer layer. Mammal teeth with enamel have evolved to be rather rugged and durable tools for chewing and macerating food.

A typical human tooth is anchored in the jawbone, and has a generalized structure with a crown and root (Figure 4.2). The crown is capped by a thick layer of enamel that sits over a layer called the dentine layer, which is less hard than the enamel layer. The dentine layer surrounds a pulpy region through which blood vessels and nerves course. Between the teeth there is a tissue layer called the gingivae that assists in anchoring the tooth above the bone line. The periodontium, which includes the cementum, is a part of the tooth anatomy that further assists to anchor the tooth in the mouth and is a layer of tissue between the bone and the tooth itself. This generalized structure means that teeth have several desirable habitats in and on which bacteria can reside. Some microbes like the enamel surface of the tooth, while others prefer the soft tissues of the dental structures such as the gingivae and the periodontal layer. Still others will

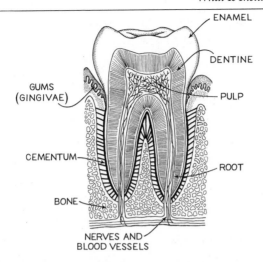

Figure 4.2. Cross-section of a typical human tooth.

colonize the pulp or the dentin layers, because they find the region around the blood vessels and nerves to be the most habitable.

Several pathologies can develop when the balance of microbes is disrupted around the teeth. One pathology with which we are all probably very familiar is tooth decay—otherwise known as cavities or dental caries. Dental caries are caused by acid attacking the teeth. Several bacteria have been associated with dental caries, specifically *Streptococcus mutans* and some species from the genus *Lactobacillus*. What is interesting, however, is that these species of bacteria are not absolute predictors of whether or not dental caries will occur. Hence, looking at the tooth microbiomes of people (usually children, because some populations of children struggle with tooth decay) has given a lot of insight into how dental caries develop in a microbial context.

Anne Tanner and her group at Harvard University have looked at this phenomenon in over seventy-five children and have shown that the extent of tooth decay is correlated with the quantity of

S. mutans and other *Streptococcus* species. Two other species that are important in another pathology called periodontitis—*Prevotella intermedia* and *Tannerella forsythia*—were found in low levels in children with dental caries. Tanner looked at children from different ethnic backgrounds and discovered that children with caries had tooth microbiomes that were very similar, indicating that the bacteria causing caries do not discriminate by ethnicity.

As we pointed out, *S. mutans* and *Lactobacillus* are not absolute predictors of dental caries. Other studies of children with extreme caries indicate that acid-producing bacteria such as *Neisseria* and *Selenomonas* are observed in high frequency when *S. mutans* and *Lactobacillus* are not involved. This result simply means that there is more than one way to produce caries. One advantage that some bacteria have is the ability to merge into a colony that "sticks around" in the form of a biofilm. Most microbes in the oral cavity are washed away when a liquid is drunk or are taken along with any food into the digestive tract, where they are more than likely killed by the high acid content of the stomach. But bacteria in biofilms have adapted to this mechanical removal strategy by banding together and producing molecules that allow the colony to adhere to surfaces such as the enamel of teeth and the skin of the gums.

Even in a healthy, clean mouth, when teeth are brushed regularly, a community of microbes will begin living on teeth almost immediately after the toothbrush is removed. Right after the teeth are cleaned or when a child's new tooth erupts, the surface of the tooth starts to become covered by a thin layer of proteins called a pellicle. This pellicle is then the substrate on which bacteria like to form films. One can look at the colonization of the tooth by microbes exactly like an ecologist would look at the colonization of a new volcanic island. Right after it erupts, the surface is sterile; then it is colonized by species that can exist under extremely limited conditions. Next, species start to colonize that specialize on interacting with the first colonizers, and succession follows with more and more complicated interactions of species. The colonization of a tooth by microbes is no different. The first microbes to attach to the pellicle are spherical bacteria also known as cocci. These bacteria include *Streptococcus mutans, Streptococcus mitis, Streptococcus sanguis, Streptococcus oralis, Rothia dentocariosa,* and *Staphylococcus epidermidis.* This first phase of colonization is followed by the attachment of several different species of bacteria, so that there are more than twenty different species cohabiting a very small layer of film above the pellicle. The film at this stage is anywhere from one to twenty cells thick.

Next, secondary colonizers add to the initial thin layer of bacteria. In addition, the bacteria that are part of the biofilm start to replicate to produce a biofilm anywhere from 100 to 300 cells thick. The different bacterial species that come in this second wave arrange themselves in layers according to their ecologies. As the biofilm increases in thickness, bacteria that are more tolerant to oxygen in the air are added, so they cover the anaerobic, or oxygen-hating, ones. Metabolism of specific bacteria of the film is also an indicator of where in the biofilm column certain species will appear. This arrangement produces a system where the anaerobic microbes live deeper in the biofilm and microbes with aerobic lifestyles are on the surface. The microbes in the film are continually communicating with each other through molecular mechanisms. And thousands of bacterial genes are producing proteins to further shore up their attachment to the tooth and integration into the biofilm.

How do the bacteria communicate with each other? They use an amazing process known as quorum sensing, which allows them to make decisions. This decision-making is not really the kind of reasoned decision-making that we humans do on a daily basis, but rather a mechanism whereby microbes can choose, so to speak, among different alternatives they have for responding to their environment. Some of the alternatives that microbes have when they are confronted with the environment are to grow in large masses or biofilms, to fight back with the hosts with whom they are associated and become virulent, or to respond to antimicrobials. There is a fine line, however. If microbes "decide" to form a biofilm before there are enough of them, the biofilm will be unsuccessful and the entire assemblage will die. Likewise, if a group of microbes starts to become virulent and attack their host but there aren't enough of them, the decision will backfire, and the host's immune system will destroy them.

Many species of bacteria have evolved quorum-sensing systems. Bacteria are continually producing a special class of molecules known as signaling molecules or autoinducers that they secrete outside the cell membrane. Meanwhile, in the bacterial cells there are other molecules called receptors that will bind the autoinducer. When binding occurs, the reaction activates a gene or a suite of genes that become important in the formation of biofilms, the development of virulence, or other mechanisms controlled by quorum sensing. In this system the bacteria don't really sense self and non-self, because the autoinducers they produce can actually activate their very own inducers. Instead, they rely on the statistics of diffusion to drive the system. If there are only a few bacteria in the vicinity, the autoinducers diffuse so much that they simply don't find inducer receptors. Only when a critical mass of bacteria is present does the system start to kick in. That is, as more and more bacteria are growing in the same area, more and more of the autoinducer is produced. When a critical concentration threshold of the autoinducer has been reached, the bacteria kick it into overdrive, and even more inducer is made. Large-scale, coordinated binding to receptors is thus initiated, leading to the cascade effect whereby genes are expressed that are relevant to a physiological property of the bacteria, such as biofilm formation or virulence.

The process of quorum sensing plays a big role in modulating how the species of bacteria in the biofilm grow and replicate. Specifically, it produces a protein called competence-stimulating protein that is important in the process whereby *S. mutans* individuals transform each other with DNA from their genomes. Transformation is an important function for bacteria, because it broadens their genetic and functional repertoire, often facilitating the transfer of pathogenic genes to harmless bacteria. When *S. mutans* is grown in small amounts in liquid medium, it uses quorum sensing to count the number of members of its species around it. When it doesn't sense a lot of the members of its own species, such as when it is in liquid media at low concentrations, it produces only a small amount of competence-stimulating protein, because if there aren't enough cells around for the protein to have an effect on other *S. mutans* in the dilute culture, then there is no need to transform. But in biofilms where there might be layer upon layer of *S. mutans,* the competence-stimulating protein in a quorum-sensing process stimulates transformation to occur as much as six-hundred-fold more than when the microbe is at lower concentrations. By sensing that there are a lot of neighbors around it, *S. mutans* can make the right protein to facilitate transformation, which in turn enhances the survival of its genome.

Although certain bacterial species have been associated with dental caries in children, the exact assemblage of bacteria that triggers the pathology is not known. Several species of *Streptococcus* are involved, and *S. mutans* in particular plays a major role in making biofilms on the teeth. *S. mutans* is also a member of the healthy tooth biofilm microbiome, so it must be the company it keeps that makes it pathogenic.

One of the traits that microbes in biofilms have is that they are copious acid producers: one of the major biochemical pathways they use is the glycolytic pathway, which ferments sugar and, as a byproduct, creates acidic compounds. These bacteria must thus be able to tolerate high concentrations of acid and a low pH. Once these bacteria form biofilms, they become very active fermenting the sugar we eat, and soon the biofilm is producing a lot of acid that is corrosive to the teeth. If the sugar intake stops, then the pH goes back up, and the tooth remineralizes. Caries are thus an indicator of an imbalance in the mineralization versus acidification processes that go on in our mouths.

How species interact with each other on oral biofilms has been studied by implanting membrane filters in the mouths of a large number of young adults in Germany. The filters were removed after one, three, five, nine, and fourteen days and washed to obtain any biofilm-creating microbes. The young people in the study were all allowed to eat their usual foods and to brush and floss their teeth as they chose. As expected, each individual had a unique pattern of bacteria living in his or her biofilm. Oddly enough, though, there were no temporal trends in the bacterial composition of the subjects of the study—that is, no patterns emerged in how the bacterial composition of the oral biofilms changed, or didn't change, over time. The most interesting result of the study was that the subjects clustered into three major groups, based on the predominant kinds of bacteria collected from their teeth. One group had a prevalence of *Streptococcus* species as previously described. The second group had more bacteria from the genus *Prevotella,* and the third had a large quantity of Proteobacteria in the biofilms. The significance of these three major kinds of biofilms is yet to be discerned, but the ability to discover major different kinds of species compositions might go a long way toward understanding the overall health of human teeth.

A very serious oral disorder called periodontitis can also occur as a result of an imbalance of the oral microbiome on the area of the gums surrounding the teeth. Periodontitis is thought to be the result of ecology gone wrong. The normal microbiome of the periodontal region basically lives in harmony with our teeth and the area around it. In fact, the health of this area of the mouth is probably dependent on the presence of certain kinds of bacteria that may fend off unpleasant ones. But there are some species, such as members of the so-called red complex, or periopathogens, that can disrupt the ecology of the periodontal region. When these species— which include *Porphyromonas gingivalis, Tannerella forsythia,* and *Treponema denticola*—are present, they use various mechanisms to short-circuit the innate system of the human host and cause periodontal disease.

Down the Hatch

After the teeth and tongue, there is the esophagus. This region of the digestive tract is important for understanding gastroesophageal reflux, or heartburn. One of the difficulties in researching the stomach, intestines, and any internal organ for that matter, is getting to those parts of the body safely and easily. The Enterotest is researchers' clever answer to the problem of noninvasively accessing the stomach and esophagus. This device is simple in design and was first developed to test the intestines for the presence of parasites. It is basically a string with a collection device at the end that the subject swallows. Once the sampling has been accomplished, the small device is recovered by pulling it out by the string. By shortening or lengthening the string, one can have the sampling device rest in different parts of the gastroesophageal tract. This is a particularly good method for sampling in small children and for adults who require deep esophageal, stomach, and intestinal biopsies. Although the research on the microbiome of the esophagus is in its infancy, it is an important endeavor because of disorders such as acid reflux and other microbially induced gastrointestinal problems.

The microbial inhabitants of the stomach have to face an extremely acidic habitat (Figure 4.3). The number and kinds of microbes in the stomach are therefore somewhat limited compared with the rest of the gut. For instance, although there are 10 billion microbial cells per milliliter (one-thousandth of a liter) in the lower part of the gut (the large intestine and colon), there are as few as ten thousand microbial cells per milliliter in the stomach. Similarly, there are tens of species of microbe that are usually found in the small intestine such as those from the genera *Escherichia, Klebisella,* and *Lactobacillus,* to name a few, and literally hundreds of species in the first category found in the large intestine. But there is only one species of bacterium that is a regular inhabitant of our stomachs: *Helicobacter pylori.* It can withstand low pH conditions and has carved out a very specific niche in the gut—a niche that is currently being disrupted in many people (see Chapter 6).

There are two major categories of microbes in the gut, depending on whether they are regular inhabitants or just interlopers. In the first category are those microbes that are usually found inside the host and commonly in a wide array of hosts. These are the expected and welcome guests that inhabit the gut. The second category of microbes is external and not found regularly in other hosts. But these unwelcome "home invaders" can become major players in any single gut microbiome by replicating copiously.

Because the stomach is separated from the small intestine by a valvelike exit called the pyloric sphincter, the ecological conditions of the stomach can change drastically as a microbe passes from the stomach into the lower gut. The small intestine is classically divided into the duodenum (adjoining the stomach), the jejunum (a little farther down, but still closer to the stomach), and the ileum (closer to the large intestine). The pH of these regions of our guts steadily increases as one gets farther away from the stomach, going from 2 or 3 in the stomach, to 7 or even 9 in the jejunum, and 7 or so in the large intestine.

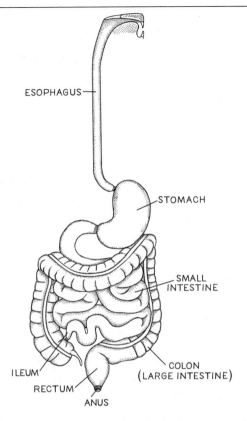

Figure 4.3. The anatomy of the human digestive tract.

As already noted, the small intestine's microbiome is generally less diverse and less populated than that of the large intestine, because the small intestine is a harsher environment for bacteria than the large intestine in two ways. First, it is infused with bile salts that are toxic to many bacteria, and it is lined with little regions of cells, called Peyer's patches, that act as sentinels for telling the immune system that something is present that shouldn't be. The Peyer's patches produce a lot of molecules from our immune system called immunoglobulin A that identify foreign cells.

The large intestine lacks these patches of cells and is larger in volume than the small intestine. In fact, the large intestine is a rather benign place—even the peristaltic action, that subtle movement of the intestines that moves food and waste along, is less strong in the large intestine. These conditions allow for bacteria to proliferate and grow to large population sizes in the large intestine. Because the large intestine also lacks certain enzymatic functions, such as the ability to absorb amino acids and B vitamins, which can be toxic in large amounts, the presence of large numbers of bacteria to absorb these chemicals and perform other useful jobs in this area of the gut is usually a good thing. In fact, the dominant bacterial species that has this important commensal function in the human gut is a microbe known as *Akkermansia muciniphila*. As its name implies, it is associated with the mucous membrane of the large intestine. These bacteria in large numbers actually regulate some very important functional aspects of the large intestine, such as the efficiency of the barrier of the intestinal wall and some metabolic functions. It is therefore a very important constituent of our guts and acts as if it is part of our digestive system. As with any other organ in our digestive system, though, if something goes wrong with it, or in this case with its interaction with our gut, problems can ensue, such as obesity and other digestive metabolic disorders.

This Way Out

The large intestine empties into the rectum, which stores and ultimately eliminates feces. There are two ways to sample the rectum—one invasive and the other less so. The invasive sampling technique employs a device called an anoscope, whereas the less invasive way is to simply sample the feces. Either way, what is obtained is a sample of the diversity of the microbial communities of the rectum and to a certain extent a sample of which microbes pass through the digestive tract.

The unprecedented focus on feces has been facilitated by the appearance of a large number of private companies and public endeavors hoping to gain insight into the fecal microbiome. Apparently there is a lot of money in the business of poop. But although this interest in human feces is new, feces in general have been of interest to researchers for a long time. Animal feces have been examined in some detail by scientists. For more than a hundred years paleontologists have studied fossil feces called coprolites and for some years now researchers have used fecal matter as a tool for genetic analysis on difficult-to-sample animals such as big cats or bears. It turns out that when a human or other animal eliminates feces, not only microbes but also cells from the rectum are sloughed off into the excrement. The DNA that is part of these cells can then be studied using the same DNA barcoding approaches mentioned in Chapter 2.

Animal feces can reveal a lot about animal identity. By analyzing the microbiome of the poop of several African monkeys, researchers were able to name their species. Red colobus monkeys, black and white colobus monkeys, and guenon monkeys are full of bacteria, but the bacteria are highly specific for each species, perhaps due to either the evolutionary separation of monkey species, their distinct diets, or most likely, a combination of the two.

Animals such as cows are sampled for microbial diversity because of their agricultural and economic importance. One can just collect fresh cow dung, but the gut microbiota of cows are so important that, to facilitate rapid access to the part of the cow's stomach called the rumen, a windowlike device called a cow cannula can be implanted in the cow's side so that samples can be taken at any time (Figure 4.4).

An important factor for the growth of cattle is diet, and by characterizing the microbiome of cattle under different diets, the cattle

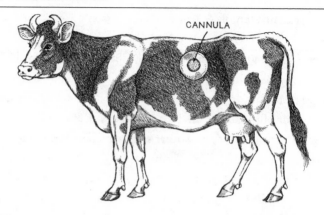

Figure 4.4. A cow with a cannula—a windowlike opening that allows researchers to have direct access to the cow's multichambered stomach.

and dairy industries have tried to correlate microbial diversity with different feeding strategies in the hopes of determining better growing conditions for cattle. Cattle are fed three basic diets: an early growing diet, a late-growing diet, and a finishing diet. The differences in the diets usually involve the percentages of dry rolled corn and silage included in the feed. Diet does make a big difference: in a study of more than four hundred cattle from Nebraska, there were an amazing 176,692 different phylotypes of bacteria in the cow pats collected, with only a little over two thousand of these consistently appearing in all three feed groups examined. The conclusion is that, amazingly, there is only a 1 percent overlap in the three feed groups, which suggests that diet does greatly influence the microbiomes of cattle and their digestive tracts, and in turn has a significant impact on the fat content and overall health of individuals. Other animals have had their microbiomes examined as well, with similar results showing a dizzyingly diverse assemblage of microbes living in animal feces.

The human fecal microbiome has been examined extensively. The most comprehensive study was done by one of the gurus of microbiome research, Jeffrey Gordon, at Washington University in Saint Louis. Gordon and his colleagues examined the fecal material of more than five hundred men, women, and children from three diverse regions of the globe: Venezuela, Malawi, and the United States. The focus was on healthy subjects, some of whom were identical or fraternal twins. The theme of differences in microbial diversity between individual people also holds for fecal material and, in turn, for gut microbiomes. But there are also regionally specific differences among the three major populations, with the U.S. microbiomes looking very different compared to the Malawi and Venezuelan ones (Figure 4.5). The differences not only show up in children, but also are evident in adulthood. What are some of the details?

First, the microbial diversity of fecal microbiomes increased with age during the first three years of life. Strikingly, children after age three in all geographic regions experienced age-associated changes in the gut microbiome. By analyzing the genes of the microbes that

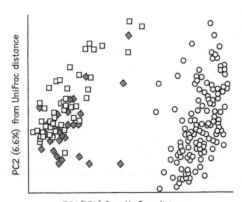

PC1(25%) from UniFrac distance

Figure 4.5. The fecal microbiomes of humans from the United States (circles) and Malawi (squares), as well as Amerindians (shaded diamonds). The x and y axes refer to sources of variation in the data; each individual point has specific values for the sources of variation and hence can be plotted as shown to indicate the similarity of the microbial communities sampled.

implemented the change, Gordon and his team determined that the bacterial genes critical to the age-associated change are involved in vitamin biosynthesis and metabolism. It is clear that the stuff that passes through our digestive tracts is susceptible to both geographic changes and age changes.

Pathogenesis of some microbes and overall health are tightly tied to what gut microbes are doing (see Chapters 5 and 6). For instance, it is now well known that obesity in mice is correlated with the abundance of specific kinds of microbes in their guts. Obese mice have one profile of microbes in their gut, whereas lean mice have another different, and distinctive profile of microbes in their gut. These differences are not diet dependent though. Mice with an obese genotype that are separated at birth and raised with others with the slim genotype have the same microbiome profiles as obese mice. This result simply means that if an individual has an obese genotype, then the composition of the gut microbiome will be dictated by that genotype.

One issue skirted so far is the viral makeup of the human microbiome. Gordon and his colleagues have focused some of their research on this subject. By looking at the fecal microbiomes of several mothers and their adult female identical twins, they determined that the virome of feces is individualized regardless of degree of relatedness. This interpersonal variation can be contrasted with the variation within a person, because the researchers also collected samples from their subjects over the course of a year. During this sampling period, a single person's fecal virome changed little. What are these viruses? They are ones that like to infect bacteria, and specifically viruses called temperate phages. Such viruses like to invade the host bacterial cell, integrate into the host chromosome, and hang out as interlopers. Most of the viral bacterial interactions studied previously focused on a different phage lifestyle called a lytic cycle. Phages engaged in the lytic cycle replicate inside the host bacterial cell and burst the cell, like popping a balloon. When bacterial cells are infected with lytic phages, the ecosystem in which they exist tends to go toward what is called Red Queen status, which refers to the character in Lewis Carroll's *Alice in Wonderland* who has to run really fast just to stay in place. Indeed what happens to interacting populations of organisms in Red Queen dynamics is that they evolve very rapidly in response to each other just to remain viable. The Red Queen hypothesis has been a focal idea for how viruses interact with cells and how bacteria interact with humans. It appears, however, that the interactions of most viruses in the gut virome do not fit this scenario. This observation has added new energy to efforts to understand why the viruses in our guts change with the bacterial hosts.

Show Me Yours, I'll Show You My . . . Microbes

The digestive tract and skin have the most diverse microbiomes, because they need microbes to deal with their everyday functions. Without the microbiomes on our skin and in our guts, we would be very different organisms—the millions of years of evolution of our guts and skin with microbes have made very sure of that. Other organs in our bodies, however, have evolved to avoid microbes. The microbiomes of these forbidden zones are less well characterized, but they do give us a good idea of what normal is, which helps for thinking about pathology and sickness.

The genitals of humans are, to say the least, very interesting. They are there for reproduction, yet we humans have figured out ways to use them for both procreation and recreation. Even when we aren't using our genitals for sex, they are swarming with microbes. But it's the recreational aspect of genitals that usually causes trouble by inviting in unwanted visitors.

Female skin microbes are somewhat different from those found on males (Chapter 3). Specifically, bacteria in the genus *Lactobacillus* appear to be abundant. It turns out that the "vagina-ome" and the "penisome" also have microbial differences. The female vaginal area has evolved to be mutualistic with a diverse array of bacteria. As with all the other microbiomes we have examined so far, there is a lot of interpersonal variation in the bacterial makeup of the vaginal cavity. But this variation clearly falls into five different kinds of communities, so that individuals with a particular dominant microbial group in their vaginal microbiome resemble each other in this way more than they do individuals with different microbial profiles in this area. These "community state types" are abundance profiles for general groups of bacterial species, with healthy human vaginas featuring constituents from the genus *Lactobacillus* or *Prevotella*. For example, Type I's vaginal microbiome consists mainly of *Lactobacillus crispatus,* Type II's is composed mostly of *Lactobacillus gasseri,* Type III's is dominated by *Lactobacillus iners,* Type IV's consists primarily of *Prevotella*, and Type V's is composed mainly of *Lactobacillus jensenii.* Of all the women surveyed, 24 percent were Type IV. All of these microbes like an environment with a low pH—in other words, they thrive in high acidity. Bacteria in the genus *Lactobacillus* actually produce molecules that precisely target the cell surfaces of other, unwelcome species and poke holes in their cell membranes. By making these narrowly effective and focused molecules, they can attack other microbes and not be harmed by their own products.

Bacterial vaginosis is a disease state of the vagina that is often confused with yeast infections and vaginal trichomoniasis (infection by *Trichomonas vaginalis*). It is caused when the normal balance of *Lactobacillus* (or *Prevotella*) is disrupted. So species in the genus *Lactobacillus* are important inhabitants of the vagina and have actually coevolved to act beneficially by preventing the growth of bacteria that could be harmful. To put it simply, the vaginal microbiome, while variable between individuals, is a delicate ecological system. If the species composition, or the habitat itself, is disrupted, then the ecological balance goes awry.

What about the penis? The literature on this topic is much more limited than that for the vagina. At the time of this writing, a search using PubMed at the National Center for Biotechnology Information brings up 230 publications for the words vagina and microbiome but only eleven for "penis and microbiome." These eleven studies, however, do reveal some interesting trends regarding the male genital tract.

Early studies of human penis microbiota used techniques that required culturing samples taken from several regions of the penis. Several areas of the male genital tract have been surveyed for their microbial diversity, such as the coronal sulcus, navicular fossa, the urethra, and the prostate. In addition, there has been some interest in microbes associated with disease states as well as in what differences there might be between circumcised and uncircumcised penises with respect to microbiome, so studies addressing these questions have been conducted using these culturing techniques. A more complete picture, however, has been discovered with the microbiome approaches described in Chapter 2, which offer much more precise analyses of species composition. To achieve large-scale microbiome analysis, researchers used a "first catch" urine sampling technique from sexually active men, then noted which men were negative or positive for sexually transmitted infections. Although there was a lot of variation between individuals' urine microbiomes, the urine microbiomes of men with sexually transmitted infections were significantly similar, as were the urine microbiomes of men without such infections. And there were very significant differences between the urine microbiomes of those who were positive for these infections and those who were not. In particular, bacteria from the genera *Sneathia, Gemella, Aerococcus, Anaerococcus, Prevotella,* and

Veillonella are usually not found in men's urine but were present in samples from men with sexually transmitted infections. The question then becomes, does the presence of these microbes allow for a sexually transmitted infection, or does the infection result in an altered ecosystem that allows these microbes to establish themselves? To date this question has not been answered.

Circumcision is a cultural practice that has been around for at least several millennia. Because circumcision involves the removal of skin from around the glans penis and frenulum, it changes the outward appearance of the penis, and, in fact, alters one of the major microbial habitats on the penis. In the flaccid penis of an uncircumcised individual, the foreskin will cover the glans and frenulum, which creates a moist, warm, and mucus-rich habitat that men with circumcised penises do not have. If one looks at the microbiome of the penises of individuals before and after circumcision, there is a difference in community makeup. Specifically, anaerobic microbes tend to be more prevalent before circumcision. By exposing the area of the penis called the coronal sulci or subpreputial space to more oxygen, circumcision produces a change in the habitat of this area of the penis. Specifically, microbes that grow in oxygen can now colonize and push out microbes that are anoxic. Clostridiales and Prevotellaceae were two inhabitants of the subpreputial space in uncircumcised men that were eliminated after circumcision.

Circumcision also eliminates a large amount of the mucus that is normally present on the penis before circumcision. The loss of the inner mucosa usually under the foreskin also causes changes in the immune system. A particular kind of immune system cell called a Langerhans cell likes mucus. In mucous membranes, Langerhans cells are activated and become cells that present antigens. If the mucus is eliminated, the number of Langerhans cells is reduced, and hence there are fewer immune cells at work. The characterization of the "penisomes" of circumcised and uncircumcised men suggests that pathogenesis associated with the penis is more complex than simple infection by a single pathogen such as HIV. It is known that male circumcision is negatively correlated with the contraction of HIV. If male circumcision has the double whammy of altering the response of the immune system and the ecology of the penis (and hence the microbiome of the penis), however, then understanding the phenomenon of HIV infection becomes much more complex.

A Breath of Not-So-Fresh Air

James Beck has said of our lungs and microbes, "The lungs of healthy humans have traditionally been considered to be sterile when examined by culture-based techniques." And so it is with the rest of the body with respect to microbes. Most of the other tissues and organs in our bodies are infused by the circulatory system. Our innate and acquired immune systems present in the circulatory system can therefore activate and keep us healthy. But as Beck points out, this is just what we know from culture-based approaches. In actuality, our lungs are home to a number of microbes and our blood, which has typically been considered hypersterile, is also rich with microbes. Our other organs are therefore also not sterile. The current view of treating disorders in these other organs as if they were caused by a pathogenic infection in an otherwise sterile space, then, is simply incorrect.

But if our organs are not sterile, what is living in them? Our lungs have contact with a diverse array of microhabitats, so almost certainly if microbes inhabit them the communities are diverse as well. In fact, although the skin is indeed the largest organ in the body, it turns out that because the lungs have thousands of little internal passageways called alveolar passages, the surface area exposed to air in the lungs is about thirty times that of the skin. The temperature of the wall of the lungs, too, can vary anywhere from 26° C to 32° C.

This is a little like being in Maine versus Miami in the summer. The air in the lungs can also vary quite a lot in temperature (from 23° C to 28° C). With respect to pH, oxygen content, and amount of gas exchange (all of which influence microbial communities), there is a great deal of variation, with the biggest range seen in the difference between the top and bottom lobes of the lung.

The lung microbiome can tell us a lot about how disease progresses in conditions such as asthma, cystic fibrosis, and chronic obstructive pulmonary disease (COPD). Characterizing the normal lung microbiome, however, has been somewhat difficult because of various technical problems. For instance, a lot of the early work on lung microbiome was obtained from sputum samples (junk coughed up from the lungs and spit into a sampling receptacle). But because sputum passes through the mouth, a certain degree of contamination occurs. Another problem is what the definition of a "healthy" lung might be. We tend to take a lot of breaths each day—nearly twenty thousand of them—and if we move about a lot during the day, we expose ourselves to a number of different types of air, not to mention pollutants like cigarette smoke.

Despite these obstacles, some initial studies have attempted to quantify a healthy lung microbiome. Some researchers have used a subtractive approach to get around the contamination issue. For instance, they will obtain both a sputum sample and a separate sample from the oral area from the same subject. If the sputum sample has microbes in it that are different from the oral cavity sample, then it is a good bet that the novel microbes are from the lungs. Others have employed a more invasive technique called bronchoalveolar lavage to reach into the lungs for a sample, though the subject has to be anesthetized for this procedure. Using these sampling procedures, researchers have found that the microbiome of the lungs resembles the upper airway, in that the major bacterial taxa present are in the phyla Firmicutes and Proteobacteria. At the genus level there are also some similarities. The big differences are in how many microbial colonies of each type exist in healthy lungs versus healthy airways.

When characterizing the lung microbiome, scientists usually compare smokers with nonsmokers. An approach called spirometry can be used to measure the efficiency with which one breathes and to tell if there is a significant departure from normal inhaling and exhaling. Spirometry measures the speed of air intake and the amount of air taken in and exhaled; these values are then charted on pneumotachographs. Perhaps surprisingly, it turns out that smokers often score quite well on these tests and hence are classified as "healthy smokers." Individuals with conditions such as pulmonary fibrosis, cystic fibrosis, COPD, or asthma, by contrast, will usually score poorly on their pneumotachographs. The other comparison often made in looking at the lung microbiome is to contrast rather drastic changes of the lung with the normal state. For instance, the microbiomes of people experiencing COPD symptoms have been compared with those of healthy people and the microbiome of a donor lung has been analyzed both before and after transplantation.

In one study conducted by John R. Erb-Downward and colleagues at the University of Michigan, Ann Arbor and the Veterans Affairs Health System in Michigan, "healthy smokers," in general, had microbiomes very similar to those of nonsmokers. And surprisingly, COPD lung microbiomes were not differentiated from the other two categories, which prompted the conclusion that there is "extensive membership overlap between the bacterial communities of the three study groups." Unlike fingertip microbiomes, lung microbiomes, even with smoking and/or COPD, are not distinct between or among individuals: instead there is a core microbiome of the lungs that consists of the following genera of bacteria: *Pseudomonas, Streptococcus, Prevotella, Fusobacterium, Haemophilus, Veillonella,* and *Porphyromonas.* In a second study, subtle differences between smokers and nonsmokers could be detected in the mouth, but again, the lung microbiomes of these two categories of healthy people did not differ significantly. It also turns out that some microbes in the lower respiratory tract that have been recovered by bronchoalveolar lavage are not found in the mouth, which means that the mouth is not the only reservoir from which the lungs obtain microbial inhabitants.

One of the main ways that ecologists test hypotheses about the effects of environment on a community is to transplant them. An organ transplant, for example, offers an opportunity to investigate a set of interesting ecological questions that ecologists look at all the time. Unfortunately, some of the most important lessons learned have to do with the human body's rejection of a transplant. Bronchiolitis obliterans syndrome, for instance, is one of the major causes or manifestations of rejection of a transplanted lung. By comparing the microbiome of the lung before and after transplant and during recovery after a transplant operation, researchers are learning about the dynamics of rejection. It is probably easy to envision that a transplanted lung provides a novel set of habitats for microbes, and this is indeed what scientists find when comparing transplanted lung microbiomes. In fact, the diversity and novelty of microbes in a transplanted lung is impressive and involves the presence of an interesting but potentially pathogenic family of bacteria called Burkholderiaceae. This family of bacteria is most famous for its role in plant pathogenesis, but certain members of the family are also human pathogens.

Now that some progress has been made in characterizing the microbiome of a normal lung, there is a lot of interest in learning how a disease state, such as cystic fibrosis, can disrupt the ecology of the lung and how this important organ reacts to such disruption.

Before we can get to these questions, however, we will explore our defenses against unwanted incursions by the microbes in and on us. In some cases, the defense happens automatically, as a result of millions of years of our coevolution with certain microbes. But in other cases, additional mechanisms of defense are needed.

CHAPTER 5

What Are Our Defenses?

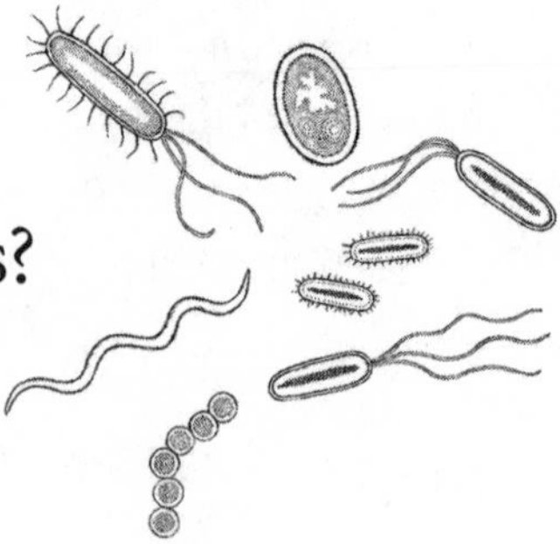

Puerperal fever is an infection caused by the bacterial contamination of a woman's reproductive tract during and after giving birth. Today, one in eight women who deliver children will contract this infection, and three in every 100,000 women still die from it. Two hundred years ago, though, the mortality rate from puerperal fever—or childbed fever as it was known then—was about a thousand times greater. In the 1840s, Ignaz Semmelweis, a physician at Vienna General Hospital, noticed that women who gave birth in hospitals and stayed overnight were much more likely to die from childbirth fever than women who gave birth at home. Why was the hospital such a dangerous place to give birth? As Semmelweis commented in his treatise on the disease, "Everything was in question; everything seemed inexplicable; everything was doubtful. Only the large number of deaths was an unquestionable reality." To Semmelweis, observation was the way to stem the tide of the disease. He first started to observe the hygienic habits of doctors, who at the time refused to wash their hands before surgery; in fact, the stiffer their operating gowns were with grime and blood, the higher respect they garnered. It wasn't until one of his colleagues died of a terrible infection after cutting himself during a routine autopsy in the hospital's morgue that Semmelweis's powers of deduction clicked. His colleague, a Professor Kolletschka, had contracted symptoms very similar to childbed fever. In a burst of inspiration, Semmelweis commented, "I was haunted by the image of Kolletschka's disease and was forced to recognize, ever more decisively, that the disease from which Kolletschka died was identical to that from which *so* many maternity patients died." Semmelweis then started to trace the disease to its common causes. He noticed that there were two clinics where women went to give birth in Vienna: one was at the medical school in Vienna (First Clinic) and the other was a clinic where midwives were trained and delivered babies (Second Clinic). The results of this examination were puzzling to Semmelweis, because the midwife clinic had far less mortality than the medical school clinic (see Figure 5.1, which includes a translation of the original manuscript).

Semmelweis reasoned that surgeons delivering babies were transferring something from the autopsy room to the delivery room that was causing the deaths in the medical school clinic. Sadly, he recognized that "only God knows the number of patients who went prematurely to their graves because of me." Semmelweis published a paper reporting his observations and was roundly criticized by the medical community of the time. In this paper he recommended that

	FIRST CLINIC			SECOND CLINIC		
	BIRTHS	DEATHS	RATE	BIRTHS	DEATHS	RATE
1839	2781	151	5.4%	2010	91	4.5%
1840	2889	267	9.5%	2073	55	2.6%

Figure 5.1. Data table from Ignaz Semmelweis's work on puerperal fever. The "first clinic" refers to the medical school of Vienna; data from the "second clinic" came from the midwife training clinic. Semmelweis's observations led to hand-washing guidelines for doctors, saving countless lives.

surgeons wash their hands in chlorinated lime solutions before doing any kind of procedure, including the delivery of babies. While his colleagues were not impressed with his ideas (his reappointment at the Vienna General Hospital was denied), Semmelweis could at least feel comfortable that his individual work had saved lives.

Joseph Lister, a British surgeon, did not read Semmelweis's work when it first came out. He did, however, read in detail the work of Louis Pasteur, which led him to the very same conclusions that Semmelweis had made. Lister noticed that phenol compounds were often placed over rotting, smelly sewage. The phenol eliminated the smell, but Lister noted that the smell had to be coming from microbes. He proceeded to develop a wash solution for surgeons that today bears his name. Listerine is nothing more than a simple compound called carbolic acid. How it worked did not matter to Lister, but it did to a chemist named Emery I. Valko and his colleague, A. S. DuBois. Valko suggested that soaps such as carbolic acid simply "narcotize" bacteria rather than kill them. Valko also showed that narcotized bacterial cells could be revived after treatment with soap. According to Valko, the positively charged carbolic acid has this narcotizing effect and prevents the bacterial cell from growing. If the positive charges are removed or neutralized, then the bacteria come back to life. Chemically what happens is that the bacterial cell wall is made of lipids with positive and negative ends. This allows the bacteria to stick to body parts such as the hands that have different charges because of the dirt, grime, or grease on them. When you wash your hands with soap, the soap both influences the cell membranes of the bacteria and loosens the hold that bacteria have on the surface of the hands, allowing the bacteria to rinse off more thoroughly. This work was so exciting that it was reported in *Time* magazine in 1942. (Some modern soaps also contain antimicrobial chemicals, and these work very differently than the classical soaps. We will return to this issue in due course.)

Vaccines

A much older method of staving off infection had its origins in India and China. This method, perhaps several millennia old, involves treating a person with material from a similar but less dangerous version of a deadly disease in order to inoculate the person against contracting the more dangerous disease later in life. Smallpox vaccination is the poster child for this approach: material from the benign cowpox disease was very successfully used to inoculate people against smallpox, a terrible killer virus. There are several other diseases that have also been treated in this way.

One of the most prolific vaccine developers was an interesting scientist from Montana, Maurice Hilleman. Raised as a strict Lutheran, Hilleman's sharp intellect even as a child pushed him toward understanding the world in a scientific context. Evolution was an important part of how he understood biology. His desire to understand the natural world in an evolutionary context drove him to read *On the Origin of Species* in, of all places, his church. He became a microbiologist, and over his illustrious career with drug companies such as E. R. Squibb and Merck, he developed more than forty vaccines. He had a keen understanding of how the human immune system works and how it recognizes microbes when they invade the body. Although the molecular and genetic mechanisms of how the immune system works would not be discovered until the last quarter of the twentieth century, he was the most prolific warrior in the war on microbes of the mid-1900s. Hilleman did not garner as much fame as Jonas Salk and Albert Sabin, who created vaccines against the devastating disease polio, but his understanding of how microbes, and even *parts* of microbes, kickstart the immune system was well advanced.

How vaccinations stimulate our immune system is an interesting story that can only be told by first describing our immune system and the immune systems of other organisms. It is also a story that is relevant to our understanding of how the microbes in our microbiomes coexist with us. It has been pointed out that it is much easier to describe what our immune systems are than what the immune system does. The human immune system is, in one word, "complex," and in two words, "very complex." To explain it in a thousand words, or ten thousand or even a hundred thousand words, is indeed a difficult task. So we will only delve into the "tip of the iceberg" of the immune system. But by taking an evolutionary approach, we can make some dent in the complexity. Perhaps the most important message is that immune systems evolved through common ancestors that had very different challenges than humans presently experience.

The first step in immunity is the recognition of self. If you are a cell and you evolve a mechanism to destroy other cells, you had better not destroy your own cells through "friendly fire." So systems that recognize self and non-self are of first and foremost importance.

How the Immune System Began

One might think that microbes have rather simple lives. With respect to the three rules by which organisms live (I run away from that, I eat that, and I mate with that), they probably spend most of their time dealing with the second rule. It certainly seems as though they have very little capacity to run away or defend themselves. After all, they are single cells with very rudimentary defense systems. But as we learned in Chapter 4, microbes can communicate with members of their own species in order to initiate coordinated responses to the environment using a process known as quorum sensing.

Consider, for example, the incredibly interesting case of bioluminescence in the light-producing organ of the Hawaiian bobtail squid. These squid have evolved a mechanism whereby they produce light from this organ for camouflage—that is, these tiny squid can hide their silhouettes from predators by controlling the amount of light that emanates from their bodies. To control the amount of light emitted, they have coevolved a symbiotic relationship with a species of bacterium called *Vibrio fischeri.* (Another species of *Vibrio, Vibrio cholerae,* is a terrible pathogen in humans causing cholera, and the story of how it became pathogenic is fascinating but beyond the scope of this chapter.) In exchange for nutrients supplied by the squid, *V. fischeri* makes light in organs that the squid has evolved called photophores. For the light to both be produced in the photophores and be bright enough to camouflage the squid, there need to be enough *V. fischeri* and they need to make the light simultaneously. In fact, only if the concentration of bacteria in the photophore is over 100 billion cells per milliliter will there be enough light produced to fool the predator. So how does *V. fischeri* count to 100 billion? The *V. fischeri* in the photophore use quorum sensing to sense the critical population size and, in turn, produce the molecule (called luciferase) that causes the bioluminescence only when it matters. Specifically, the bacteria make small amounts of an autoinducer, and only when the photophore attains a critical population size of 100 billion is a threshold reached. At this point the autoinducer starts to bind to receptors that trigger the synthesis of luciferase by genes in the *V. fischeri* genome and, voilà, there is light! Light that benefits both the squid and the bacteria.

Bacteria need to communicate with each other, and they do so by using molecules. They are also made up of molecules that have evolved over billions of years, and these molecules look quite specific compared with the molecules we use in our bodies. Like quorum-sensing components, these other molecules are designed specifically to do the job the bacteria needs and so they look very foreign to the eukaryotic cell. It is this foreignness that the immune systems of animals, and—believe it or not—plants, use to recognize self from non-self.

The Body's Defense System: A Bacterium's Experience

Pretend you are an infectious bacterium. What would you encounter as you moved into a new home in a human body? There are many routes into the body through which you can enter, and some of your infectious colleagues have developed specialized ways of entering the bodies of their hosts. Some will hide out in food or liquids and try to enter when the host eats. Still others lurk around the genitals and in the fluids that are part of their host's sex act. Others enter through cuts or simply get into the mouth and find a soft tissue to colonize. And some will simply reside in the crevices of the body, such as the belly button, or the armpit, or between the toes. For this exercise, imagine you are one of those airborne bacteria that have specialized ways of entering the body through the respiratory tract. You float in the air, waiting to be sucked in by the inhalation of your victim. Your goal as an airborne bacterium is to get to an appropriate tissue such as the lungs, find a nice place to stick, establish yourself, and divide to produce offspring with your genome in them.

You hope you don't encounter skin, because in animals it is the first real line of defense. Skin has evolved to be very exclusive as to what it lets in and out. Once you encounter skin, your goal will be thwarted. It is dry and inhospitable. But as you cruise by, you notice there are a lot of other species of bacteria happily living on the skin. You luck out and miss the skin, and are sucked into the nasal passage, where you encounter a bunch of hairs that threaten to knock you away from your goal. You manage to dodge them, then hurtle through the nasal passage and on to the lungs. There you land in one of the many tiny passageways of the lungs called an alveolus, where you encounter a landing strip lined with mucus. You might have defenses against some of the chemicals being secreted into the mucus, and if you get past them you alight on the external tissue of the alveoli. You establish yourself and start to make proteins, an activity that makes it clear to the host cells that your intention is to stay. This is when your host's body starts to respond at the cellular level—rapidly and with a vengeance—with what is called its innate immune system. The response is vicious and well planned, because it is the product of millions of years of evolution.

You, the bacterium, have made it as far as the lungs. But as you sit in the alveoli trying to establish yourself, a cell about one hundred times larger than you stalks you with every intention of engulfing you. This cell, known as a neutrophil, is incredibly persistent, because it can sense your presence. It is part of the innate immune system of the vertebrate you have infected, and its job is to consume bacteria like you through a process called phagocytosis. Many kinds of cells other than the immune cells in vertebrates can phagocytose. In fact, some single-celled organisms make their living by engulfing other living microbes. As we pointed out in Chapter 1, two of the eukaryotic organelles, mitochondria and chloroplasts, are the result of early eukaryotic cells having engulfed bacterial cells.

The human immune system, however, has a more complicated system for phagocytosing invading material, because the cells involved are very selective about what they eat. The bacterium you represent secretes chemicals called chemokines that leave a trail of "bacterial stink" that a phagocytic cell can follow due to its chemokine receptors. (There are some great videos on the web showing phagocytic cells of the human immune system persistently chasing down invading bacteria; see References and Further Reading.) The chemokine receptors of the neutrophil chasing you, the bacterium, have been produced by genes in the genomes of the host you have invaded. There is a whole family of these receptors, which makes them capable of organizing a lethal response to a wide variety of microbial threats. But how did this innate system we describe in your human host come to be so specialized and successful?

Green Immunity

Plants are our most distant multicellular-organism relatives on the planet. They are even less related to us than fungi (mushrooms and yeast). Yet they have evolved immune systems that can recognize self from non-self, and they too will try to destroy anything they sense as non-self. In addition, plants can mount very specific responses to the many bacterial species that come into contact with them and attempt to invade them, and they have an immune memory of some invaders. One of the major differences between human bodies and the bodies of plants is that plants can communicate with distal parts of their bodies only through a simple vascular system; they don't have a highly organized circulatory system. A circulatory system allows animals to send molecules all around their bodies and enhances the rapid response to challenges our bodies face from infection. But without a circulatory system how do plants defend themselves?

When a pathogenic microbe encounters a plant, it is first faced with the plant's tough cell walls, which can be fortified with specific molecules to enhance defense. This initial fortification response happens when the plant cells recognize bacterial molecules such as flagellin (a protein important in bacterial flagella) or other molecules that are specific to bacterial cells. They do this by sensing the bacterial molecules and binding them to receptors called pattern-recognition receptors. Plant pattern-recognition receptors are actually similar structurally to animal pattern-recognizing molecules in the animal immune system response, such as Toll-like receptors. But most researchers suggest that the similarity in structure is convergent—that is, evolution in each lineage (plants and animals) has independently resulted in a similar and very efficient molecular plan for these receptors; neither lineage influenced the other.

These bound receptors trigger the production of resistance proteins (also known as R proteins), which plants then use to battle any infectious agents. The whole process is called a hypersensitive response, because it acts by destroying or causing the death of the infected host plant cells and by dispersing molecules that kill any microbes near those infected plant cells. Although this system is quite different in many ways from the processes that animal cells initially use to recognize infections, researchers have also called this an "innate" immune response.

Immunity in Lower Animals

When a neutrophil chases a pathogenic bacterium, it ignores other bacteria that it has identified as harmless. By looking at the immune systems of animals at the very base of the animal branch of the tree of life, we can understand why the neutrophil is selective in its pursuit. Lower animals have innate immunity, too. The common ancestor of multicellular animal life lived in a pretty harsh world 600 to 700 million years ago. Bacteria and other single-celled microbes had been around for a long time, and these new multicellular organisms, while they shared common ancestry with microbes, had to contend with a multitude of invaders and "ne'er-do-wells." And although multicellular animals shared common ancestry with plants at some point in the evolutionary past, plants evolved their innate system of immunity after diverging from this common animal ancestor, which was left to develop its own strategies.

Although the evolution of the earliest multicellular animals, or metazoans, is controversial, with a search for the "mother" of all metazoa engaging researchers for the past decade, scientists do know which animals are at the base of the animal tree of life. By tracing the evolution of genes that provide extant lower animals' immunity to infections from microbes, we can reconstruct the steps that were involved in molding our own immune response to microbial invaders.

Hydrozoans are bizarre-looking organisms from this lower part of the tree of life where a lot of research attention has been focused. They are members of a larger group of animals collectively called Cnidaria that also includes corals, jellyfish, and anemones. *Hydra* has a top end (logically called a head) and a bottom end (equally logically called a foot). It also has two cell layers—an endoderm and an ectoderm—as opposed to most other animals (including us) that have three (Figure 5.2). Because microbes in general colonize cell layers, it is important to keep track of these cell layers. There are also several related species in the genus *Hydra* that researchers use to study lower animal biology. Hydrozoans are very simple animals without a brain but with a nervous system called a neural net.

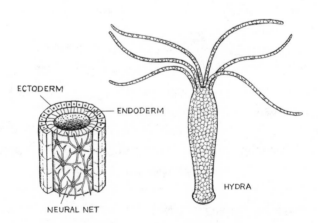

Figure 5.2. The two cell layers and neural net of a hydra (*left*), and the entire organism (*right*).

Lower animals such as hydrozoans have evolved an innate immune system, albeit a limited one. Like plants, they have pattern-recognition receptors that sense the presence of molecules made by foreign organisms, mostly microbes. In *Hydra* these are called microbe-associated molecular patterns, or MAMPs. This system tries to recognize, as a way of sensing danger, molecules that bacteria make that are not made by the host hydrozoan. If a molecule like a bacterial heat-shock protein is floating around because bacteria are near, then one of the MAMP receptors in the hydrozoan innate immune system will bind to the heat-shock protein, and its binding will trigger two responses in the hydrozoan. First, the infected hydrozoan will make antimicrobial molecules that then attack the invader, and second, it will tell the infected cell to die. This latter form of cell suicide is called apoptosis, and while it seems cruel, it performs an important service for the rest of the hydrozoan body. There are two kinds of receptors that perform the innate immunity response in hydrozoa. The first, which we mentioned earlier, are called Toll-like receptors, whereas the second are called NOD-like receptors (for nucleotide-binding oligomerization domain–like receptors). Genes for both of these receptors also exist in higher-animal genomes and indeed are utilized in both the invertebrate and vertebrate innate immune response.

This innate immune system is an efficient way to protect the body from general invasion by microbes. But does it serve as a castle wall to keep out any and all foreigners? Thomas Bosch, a hydrozoan researcher, has suggested that the innate immune system is more than just a great barrier. By looking at the microbes associated with several species in the genus *Hydra,* he and his colleagues recognized that different species have different communities of microbes associated with them (Figure 5.3). In fact, the species of bacteria in these communities were very specific with respect to the different

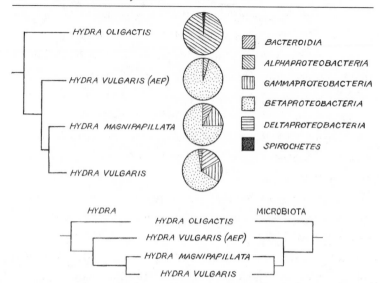

Figure 5.3. A phylogenetic tree showing the relationships of four hydroids is in the upper panel. The microbial community living in the various hydroid species is shown in the pie diagrams next to the species name in the top panel. The bottom panel shows the phylogenetic tree for the hydroid placed next to a network showing the similarity of the microbiomes from the four species of hydroid. Note that the two networks are identical.

hydrozoan species. These results are important because they show that the association of microbes with their hydrozoan hosts is very specific. Another way to put this is that the microbes seem to have coevolved with their hosts. If microbes and *Hydra* have coevolved, then perhaps some of the microbial species are really important to the lifestyle and survival of the hydrozoan.

What this implies is that some microbes actually need to be allowed past the innate immune system. In fact, perhaps our innate immune system evolved not to primarily prohibit infections, but rather to keep in proper balance those microbes that are beneficial to the existence of organisms.

Even if an invading bacterium shakes off the neutrophil, there are other molecules that can, and will, mount a defense. A system called the RNAi pathway, discovered in the late 1990s in *Drosophila melanogaster* (the fruit fly), is an important, innate defense mechanism that other animals and even plants have. Although the specifics of this and the other three pathways in *Drosophila* are beyond the scope of this book, a closer look at their basic mechanisms can start the discussion of how foreign molecular patterns trigger specific downstream responses.

The RNAi pathway acts like a Veg-O-Matic, one of the first food-processing appliances sold on television in the United States in the 1960s. If a foreign virus is detected by the system, a protein called Dicer cleaves the viral RNA. This RNA is then incorporated into a larger protein complex called a silencing complex, which then binds to viral RNA genomes. Next the silencing compound is recognized by a degrading system that then destroys that particular viral RNA genome (and anything else attached to that silencing compound). The other three pathways—Toll, Imd, and Jak-STAT—are called signaling pathways. When the receptors in these pathways notice molecular patterns specific to invading microbes and bind to those microbes, they also turn on, like a switch, the expression of antimicrobial genes in *Drosophila*. Some of these genes have names like cecropin A, drosomycin, defensin, drosocin, diptericin, and attacin A. Each of these antimicrobials, and indeed all innate system antimicrobials of animals, have different shapes and will attack different kinds of viruses, bacteria, or fungi. But all in all, their range of attack is relatively nonspecific.

Let's return to the imaginary bacterial attack. While some of the molecules start to attach to the cell surface of the bacterium, still another defense system, the complement system, is activated. This system is best described as a cascade and is also present in the lower animals and plants we have discussed. It is called a complement system because it works in concert with, or complements, part of the acquired immune system (described later).

The complement system works by using one or more of twenty proteins to bind to the molecules that have attached to the cell surface. (The complement protein can also attach directly to sugary molecules called carbohydrates that stick out of the cell membrane.) Once the complement protein binds to the cell surface, a very rapid and dynamic response is initiated. The complement molecules belong to a family of proteins called proteases that, as their name implies, degrade the bacterial proteins. The protease activity of the complement protein is only activated when it is bound to a carbohydrate or to another immune molecule. In a brutal chain reaction for the bacterium or other microbe, the initial activation of a complement protein induces the activation of another, and another, and another complement molecule. This cascade could eventually lead to the microbe being coated with complement molecules that then guide other immune cells to its location. The complement proteins can also kill the microbes themselves, by poking holes in the microbes' plasma membranes.

The host cells (neutrophils and other blood-borne immune cells called macrophages) that have been attacking the microbe come from a single progenitor cell type in the host body. The human body—and indeed any vertebrate body—has hundreds of specialized cell types, whereas some of the organisms we have discussed, such as Placozoa and sponges, are very limited in their cell type repertoire (Figure 5.4). How cell types develop or arise from a single fertilized egg is a fascinating story, and how the specialized cells of the immune system arise is simply a subplot of the general way that cells differentiate. Human immune cells, as well as the cells of the vertebrate circulatory system, arise from the same progenitor cell type called a multipotent hematopoietic stem cell. This is just a

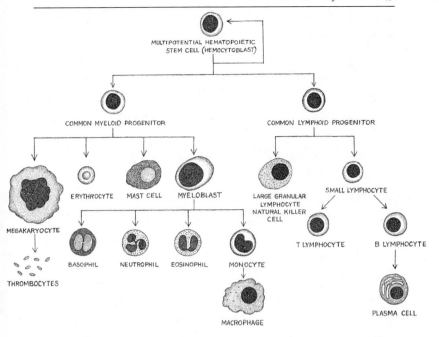

Figure 5.4. A "family tree" of cells in the vertebrate circulatory system, showing in part that all cells in the system come from a progenitor cell called a multipotential hematopoietic stem cell, or hemocytoblast.

fancy name for a cell that can turn into any of the fifteen or so cell types in the vertebrate blood system and immune system, by differentiating again and again, in a branching manner, until each of the types is created. The first differentiation determines whether it will be a lymphoid cell or a myeloid cell. The myeloid cells produce a broad variety of cell types, including our red blood cells and some of the cells we have already discussed with respect to the innate immune system.

Some of the myeloid cells produced by the multipotent hematopoietic stem cells are called leukocytes, which are the phagocytic munchers of the human innate immune system. Most of the leukocytes in our bodies are the neutrophils we encountered earlier.

They are voracious eaters of invading cells and foreign debris, and accomplished wanderers to boot. When part of the body is infected—such as when a person has a cut invaded by some bacteria—pain, heat, redness, and swelling occur. The injured cells being attacked by bacteria or viruses also start to produce a variety of chemicals including interleukins (that communicate with macrophages), chemokines (involved in recognizing chemicals in the blood stream), and interferons (antiviral responses). The interleukins and chemokines signal to the neutrophils that dinner is served. The neutrophils will then move toward the infected sites via the bloodstream. And if a virus has infected a cell, the release of interferons will signal to the infected cell that it is time to die.

If the unwelcome microbe is a virus, a natural killer cell will be summoned by the body. This kind of cell is important in the innate immune response and uses a strange strategy to fend off infection. Whereas it will produce some nasty proteins and enzymes that can knock you out, its major focus is to target sick host cells and induce them to commit suicide (or as an immunologist would say, it induces apoptosis). It is important to note that the natural killer cell leaves uninfected host cells alone. It can do this because the normal, uninfected cells put molecules on their surface called major histocompatibility complex class I (MHCI) markers. These markers signal to the natural killer cell to halt and move onto the next cell. There is a second class of major histocompatibility complex molecules called MHCII that are important in the acquired immune system. But for now, let's assume that one nasty bacterium has managed to elude all of these innate defenses. What happens next?

The Acquired Immune System

With the evolution of vertebrates in the last 500 million years, along came a second way to combat invading microbes. This more derived immune response is much more specific with respect to what it targets and how it works. Although there is some evidence that the innate immune systems are pretty flexible, the innate system doesn't really remember which microbes and other foreign bodies it has encountered and destroyed. The innate system starts from scratch pretty much every time it is confronted with something foreign. A more efficient mode of staving off invasion (or, as the *Hydra* example suggests, letting the good guys in) would involve the immune system's remembering what it had encountered in the past, and whether the encounter was good or bad. Such a sophisticated system has indeed evolved in vertebrates. Called the acquired or adaptive immune system, it appears to have evolved in the common ancestor of all jawed vertebrates.

Adding a second dimension to the vertebrate immune system introduces even more complexity to the story. If the two immune systems, innate and adaptive, worked independently, then explaining these processes would be pretty easy. We would simply have two different stories. When the adaptive system evolved, however, it did so in concert with the innate system, so there is considerable overlap of the two systems. To understand how this line of defense works, we need to know which cells are involved in the acquired response, and where they come from.

Three cell types that haven't yet been mentioned but are essential to the acquired or adaptive immune response are B-lymphocytes, killer T cells, and helper T cells. All of these lymphoid cells (a group that also includes natural killer cells, and B and T lymphocytes) are produced in generalized tissue sources in the body called hematopoietic tissue and take up residence in specific parts of the body. The myeloid precursor cells reside mostly in the bone marrow, while the lymphoid tissues reside in the lymph nodes, spleen, and thymus gland, and, surprisingly, in the mucosa of the digestive and respiratory tracts. In a complex dance of interactions with major histocompatibility complex molecules, specialized molecules called immunoglobulins and each of the other three cell types make up the acquired immune response.

The remarkably specific and effective response of B lymphocyte cells demonstrates well this subtle dance. B lymphocyte cells are covered with molecules called antigen receptors, which are ready to attach to an antigen—that is, a foreign molecule or part of a molecule, or cellular debris from something foreign. So far this sounds a little like the innate immune system with its receptors for foreign material. But there's a twist to this part of the story: when an antigen encounters the antigen receptor on the B cell surface, the B cell starts to produce more receptors and secretes them into the extracellular space. These free-floating receptors are called antibodies or immunoglobulins (Figure 5.5). Moreover, B cells remember previous infections and are able to mount a more robust and deadly attack on antigens that they have encountered earlier. As B cells replicate, they form lineages of antibody genes based in part on the antigens they

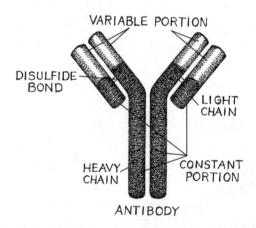

Figure 5.5. A typical antibody. Note that some parts are constant, or conserved, whereas other parts can be changed to adapt to new challenges posed by microbes.

remember. These antibody genes can then recombine in different ways to produce a large number of differently tinkered final products, each designed to attack a kind of antigen. (T cells, which also use antibodies to recognize foreign proteins or antigens, are important to how the B cell is activated to produce a specific antibody.) The antibody needs to have a conserved aspect to it, but also to cope with the large variety of microbes that invade the host body. They furthermore need a variable region that can specialize in binding to molecules of invading pathogens.

Antibodies stick really well to microbes because they act like a lock and key with some of the proteins on the microbe's cell surface. Even if you, the microbe, produce a toxin in a last-ditch effort to harm the host organism, other antibodies have the ability to interlock with that toxin, disabling it. At this point you, as the microbe, are completely coated with antibodies, and have become an attractive meal for scavenger cells in the area. Your life as a microbe has ended, you are engulfed by a voracious white blood cell and die.

The immune system is a pretty amazing evolutionary adaptation. It learns and has a fantastic memory. When it is working well, it allows certain bacteria that are beneficial to us take up residence in and on us, while attacking bacteria that are harmful. It can also be trained to ward off bacteria and viruses even before they infect us. The vaccinations we discussed are a good example of this preemptive strike capability. Maurice Hilleman created his forty or so vaccines by presenting to the human body bits and pieces of microbes that are pathogenic to us when they are alive and whole. He minced up the bacteria or viruses and introduced them to the human immune system, whose lymphocytes recognized them as foreign, and whose ensuing cascade of T cell and B cell interactions activated B cells to produce antibodies. If that pathogenic microbe invaded the body later on, those B cells would remember the pathogen and quickly produce large quantities of the antibody to combat it. The time saved by having the B cells ready for that particular infection could be enough to stop that infection before it replicated to a point where it could start making us sick or get out of control.

More than one strategy has been developed across the animal and plant world to deal with microbial infection. We have already seen that the innate systems of plants and animals both provide the first line of cellular defense but that these organisms use different molecules for this task. In addition, natural-killer cells appear to work differently in mice and men. In the mouse, the natural-killer cells use receptors called lectin receptors, whereas in men (and women), the natural-killer cells use antibodylike receptors. Jawless vertebrates like the hagfish and the lamprey, too, use very different antigen-recognition receptors in their adaptive immune system. The human kind of adaptive immune system is found only in jawed vertebrates.

What Does It Mean to Be Antimicrobial?

Legend has it that the first scientifically developed antimicrobial was the result of slacking off. Alexander Fleming, a Scottish microbiologist, returned to his lab in London from a summer vacation in the country to find that he had forgotten to discard some Petri dishes. Before he left on holiday he had plated out some bacteria in the genus *Staphylococcus* on an agar plate. He had finished his observations, but then in his hurry to start his vacation, he left the Petri dish on his lab bench without cleaning up. While he was enjoying the countryside, the plates began to get a little moldy. The fungi had most likely gotten there because, as he was observing the staph on the plate before the vacation, he had lifted the cover of the plate and an airborne fungal spore had drifted in. As hard as any good microbiologist might try, this happens frequently.

Fleming coyly described the morning after his vacation: "When I woke up just after dawn on September 28, 1928, I certainly didn't plan to revolutionize all medicine by discovering the world's first antibiotic, or bacteria killer, but I suppose that was exactly what I did." When he examined the plates on his bench he discovered that the staph colonies growing far away from the moldy blotch were nice and robust, but those growing near the mold were either dead or growing very poorly. He reasoned that the mold was somehow killing the staph, but he also knew that the only way it could be happening was through something the mold was secreting into the agar around its colony. After identifying the mold as a species in the genus *Penicillium* (Figure 5.6), he grew it in pure culture and isolated the stuff it secreted, first calling it "mold juice," then renaming it penicillin (Figure 5.7). This discovery was the start of learning how antimicrobial substances like penicillin work.

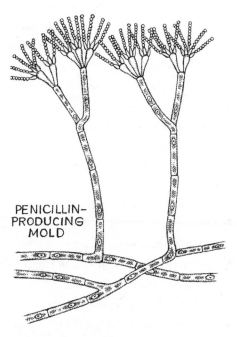

PENICILLIN-
PRODUCING
MOLD

Figure 5.6. A typical Penicillium mold.

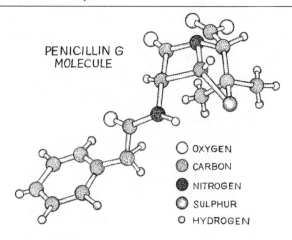

PENICILLIN G
MOLECULE

○ OXYGEN
◉ CARBON
● NITROGEN
◍ SULPHUR
○ HYDROGEN

Figure 5.7. Molecular structure of penicillin G, a form of the antibiotic administered intravenously.

By purifying and determining the structure of penicillin from the fungal strain that Fleming first isolated, chemists were able to determine its structure and to make it a viable commercial antimicrobial (Figure 5.7). Penicillin is a relatively small molecule that binds to and inactivates an important bacterial protein called a transpeptidase. It does this by slipping into the transpeptidase enzyme's active site and causing the enzyme to self-destruct. Transpeptidase is vital to the bacterial cell because it is the protein that allows it to form cell walls. Cells that are susceptible to penicillin die quickly because they cannot reproduce. In fact, because penicillin-susceptible bacteria go through all of the motions of dividing but simply can't apportion the components of the two daughter cells into real cells, they fill up and explode. There are many penicillin derivatives that have small differences in their molecular structure compared to the original antimicrobial, but they all work on the same general principle.

When we are sick, doctors can prescribe antimicrobials for the infections that bother us. There are a lot of different kinds of antimicrobial medications to use. One researcher who focused his career on development of antimicrobial substances was the New Jersey scientist Selman Waksman. He, along with his students and assistants, developed twenty-two different antimicrobials, including streptomycin and neomycin. He won a Nobel Prize for his long and illustrious career. The theme on which he worked so diligently was that certain compounds could interfere with the everyday lives of bacteria. They might impede the construction of the bacterium's cell wall, as is the case for penicillin; interfere with the way proteins are translated, as with streptomycin; or disrupt the bacterium's gene-transcription processes.

Whichever way these drugs work, they all utilize pretty much the same general mechanisms, which leads to the problem of "resistance." Bacteria and viruses can evolve to become resistant, that is, to survive the action of antimicrobial and antiviral agents. Although resistance originates as changes in single viruses or bacteria, it is important to keep in mind that antimicrobial resistance and antiviral resistance are population-level phenomena. Such resistance can happen very rapidly, depending on the population of bacteria or virus involved. HIV, for instance, can evolve resistance over a period of weeks, or even days.

The Importance of Variation

Charles Darwin was infatuated with variation. In his *On the Origin of Species,* he expounds on the astonishing variety of organisms in the natural world, using more than three thousand words to describe variation in pigeons alone. His obsession with variation, however, makes sense given how central it is to his idea that rocked the world—natural selection. He made it clear that, without variation, natural selection would not proceed. He tried very hard to connect the idea of variation with inheritance, which was appropriate. Although he did not succeed in this effort during his lifetime, when the work of

Gregor Mendel was rediscovered and the two great laws of genetics were reformulated, inheritance was indeed tied to natural selection.

Where does this variation come from? At the most basic level, mutation of the genetic material is at work. In fact, part of the reason that microbial resistance in populations occurs so rapidly has to do with how frequently the genetic material mutates to produce new variants for natural selection. The other part concerns how intense the selective pressure is for or against the new variants produced by that particular mutation. RNA viruses mutate very rapidly—at a rate about 100,000 times that of eukaryotes and single-celled microbes. And for HIV, about one base in every ten mutates in its genome each generation. In a population of millions of HIVs, this means that there is ample variation for natural selection to work, especially because some variants are more successful at reproducing than others. The reason one variant might be more successful than another with respect to selection can be understood by thinking about what an antiviral or antimicrobial does. If an antimicrobial is sprayed onto a growing population of bacteria, and all of the cells there are susceptible except for one with a mutation, then the only cell left after the antimicrobial treatment will be the one with the variant.

This means, too, that the rate of resistance also depends on how well the antimicrobial or antiviral works. In a population of thousands of viruses where the efficiency for the selection of a variant is one in a thousand, the frequency of the favorable variant will increase very slowly, but steadily. In a population where the selection for a variant is one in ten, or an even larger proportion, the frequency of the variant will increase rapidly. And if the variant is the only genotype that survives, then the frequency of that variant will explode.

For an organism or a virus to respond rapidly to an environmental challenge, genetic variation is needed. Likewise, for multicellular organisms to be able to respond as a population to the threat of an antimicrobial agent, genetic variation has to be present, because it can lead to natural selection. Several mechanisms have evolved to answer this need for variation. Bacteria have mutation rates that are about the same as eukaryotes, which have a rather efficient mechanism for increasing genetic variation—sex. With eukaryotic sex, two genomes come together, and when new sperm or eggs are formed during meiosis, the two genomes can recombine by means of "crossing over," a very complex molecular mechanism whereby two strands of DNA exchange genetic material. Eukaryotic sex at the molecular level is simply the mixing of genes from two different organisms via a fertilization process. Using sex, eukaryotes can generate pretty variable genomes, which enable populations to respond to natural selection. Bacteria do not have sex, and hence they have evolved other mechanisms, in addition to mutation, for generating variability, all of which can be placed under the general umbrella term "horizontal gene transfer" (Figure 5.8).

There are three major mechanisms by which horizontal gene transfer can occur. The first is through *transformation.* Because there are DNA strands floating around everywhere, this exogenous DNA can sometimes come into contact with intact bacterial cells. The process of transformation is a relatively sloppy one, whereby the exogenous DNA is sucked into a bacterial cell and then either coexists with the bacteria or gets integrated into the bacterial chromosome. Another mechanism is *transduction,* that is implemented through the infection of bacterial cells with viruses called phages. In this process, foreign, viral DNA is transduced to a bacterial cell via a phage that binds itself to the cell's surface. This DNA then becomes incorporated into the bacterial chromosome, where it rests in what is called a lysogenic state. The virus does not have the genes for making the cellular machinery with which it will replicate, so it relies on

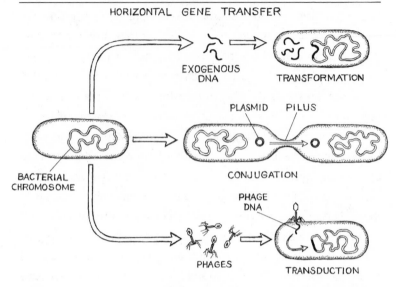

Figure 5.8. Three of the major modes of horizontal gene transfer—transformation, conjugation, and transduction.

the machinery of the bacterial host. When the time is right for the virus to replicate, it extracts itself from the bacterial chromosome and then "goes viral" by replicating, reconstituting its coat and tail, and then moving on, in the process possibly even killing the host cell by blowing it apart or lysing it. Often the extraction of the phage from the host bacteria is not exact and some of the host bacterial DNA comes along for the ride. When the virus infects a new bacterial host and integrates into its genome, these genes, or even entire suites of genes, can be a "hostess gift" to that cell. If the genes convey a reproductive advantage to the new bacterial host, that host will keep and use them.

The third mechanism, called *conjugation,* involves the transfer of small, endogenous, circular pieces of DNA from one cell to another. The transfer is made by what is called a pilus, which is simply a structural conduit through which DNA is shuttled. Circular DNAs, called plasmids, can carry genes on them, and they act like internal parasites (much like phages), but they do not integrate into the bacterial chromosome. They are parasitic though, because they rely entirely on the host's replication, transcription, and translation machinery to make copies of themselves. Sometimes the parasitic relationship turns "mutualistic," so that the plasmid and the host cell both benefit from the interaction. This occurs when the plasmid carries a gene or set of genes that increases the reproductive success of the bacterium, like an antimicrobial resistance gene. The plasmid benefits, because it can make copies of itself, while the microbe's situation is also improved, because it gains resistance to some antimicrobial compound.

Plasmids have been used for decades now in molecular genetics experiments, because they are easy to isolate and to manipulate in bacterial cells. But although plasmids are in general rather small and carry simple genetic repertoires, they can also carry loads of genes that have various functions. The smallest are on the order of a thousand nucleotides and carry genes for resistance to kanamycin and other antimicrobials. Some of the largest plasmids, by contrast, are nearly half the size of an average bacterial chromosome and carry hundreds of genes. Because they are an efficient way to generate variation in microbial populations, some plasmids carrying antimicrobial resistance genes have become quite a challenge to those trying to control modern infectious diseases like *Salmonella enterica serotype typhimurium (S. typhimurium)*, an intestinal pathogen that is facing off with a plasmid that confers resistance to ampicillin, chloramphenicol, streptomycin, sulfonamides, and tetracycline, all commonly prescribed remedies. Another particularly nasty set of resistances conferred by a plasmid is methicillin-resistant *Staphylococcus aureus,* or MRSA. This plasmid confers

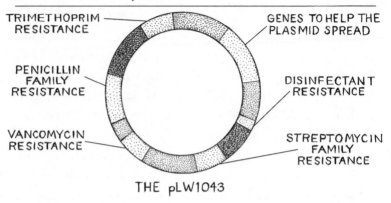

TRIMETHOPRIM RESISTANCE

GENES TO HELP THE PLASMID SPREAD

PENICILLIN FAMILY RESISTANCE

DISINFECTANT RESISTANCE

VANCOMYCIN RESISTANCE

STREPTOMYCIN FAMILY RESISTANCE

THE pLW1043

Figure 5.9. The plasmid pLW1043, of *Staphylococcus aureus,* which can confer resistance to many different antibiotics.

resistance to a number of antimicrobial agents and is particularly worrisome because it was one of the first to show resistance to vancomycin, the "go-to" drug for many infections (Figure 5.9). The plasmid of MRSA seems to have stacked the deck by collecting genes that ensure its survival and in turn the survival of staph strains with the plasmid.

The dangers of bacterial resistance and adaptation have prompted a great deal of research on why bacteria are so successful at moving genes around. One major discovery is that groups of genes can act a lot like floating islands. These so-called pathogenicity islands (PGIs), which are sometimes composed of large chunks of DNA, can be carried by phage or by plasmids, or they can be autonomous. They usually have suites of genes that enhance their ability to infect organisms or that improve their odds of survival. One example is the "TAd island," which is pretty widespread across a lot of bacteria. TAd stands for "tight adherence," and the TAd PGI is chock-full of genes that assist in making biofilms. We will return to this and other resistance mechanisms in Chapter 6, because they are extremely important for understanding the dynamics of how to maintain human health in populations that use antimicrobial agents.

So far we have looked at our immune system, vaccines, and antimicrobials. But there is yet another way to defend against pathogenic microbes: altering our own ecology.

The Great Camel Dung Mystery

When the Nazi army invaded North Africa in 1941, the German tank drivers thought it was good luck to run over piles of camel dung. Little did they know that camel dung would become a life and death issue. The Allies started to make fake camel dung piles and connect them to explosives that would detonate when run over by any luck-seeking tank. The deceit was so well planned that the Allies even put tire track marks in their fake dung piles to trick the tank drivers into plowing over them. But camel dung in its true form would hold a life-saving secret. The soldiers were suffering greatly from dysentery and the Nazi medical corps was brought in to attempt to alleviate the outbreaks by figuring out which indigenous microbe from water or food was causing the problem. Early on the local nomads were thought to hold a key to the solution, because dysentery was very rarely a factor in their morbidity. In fact, when an outbreak of dysentery occurred, or even when slight diarrhea was experienced, the nomads would diligently follow their camels around. When a camel defecated, the nomad would quickly scoop up the dung and ingest some while it was still steaming. Only recently defecated dung would work, because cold dung would not prevent the dysentery. The medical corps knew that the dysentery they were dealing with was more than likely caused by a microbe, and after close scrutiny of the dung, the corps discovered that the dung was loaded with the bacterium *Bacillus subtilis*. (This species of bacterium is in the same genus as a terribly pathogenic species, *Bacillus anthracis,* which causes anthrax, an often lethal respiratory disease. *Bacillus subtilis,* however, has since become one of those bacterial species considered "good" for humans.) What is it about *B. subtilis* that makes it so useful that Arab nomads would ingest camel dung? This species is a voracious eater of viruses and other bacteria, and it essentially clears out any and nearly all bacteria in the gut once it gets there. By ingesting the warm camel dung, the nomads were essentially altering their gut ecology to get rid of the pathogen causing the dysentery. The *B. subtilis* was present only in the warm dung; it would die out when the dung cooled. Not wanting the troops to ingest camel dung, the German high command and medical corps instead cultured large amounts of *B. subtilis* in vats and fed the broth from the cultures to the troops, stopping the outbreaks of dysentery. The Nazi medical corps even developed a way to dry out the *B. subtilis* and put it into powder form for their troops. Since the Nazi experience with camel dung, *B. subtilis* has been used in much the same way as an antidysenteric agent.

Many animals also eat their own dung. Baby rabbits are notorious for imbibing their mother's fecal pellets, and this behavior has been described as adaptive for the babies, because it is supposed to be the way they obtain their core gut microbiome. Specifically, baby rabbits need to get rid of the bacteria from the family Bacteroidaceae that they obtain shortly after birth and replace it in their digestive tracts with bacteria from the phylum Firmicutes, and the families Lachnospiraceae and Ruminococcaceae. If baby rabbits are prevented from eating their mothers' poo, their digestive tract microbiomes are disrupted, which most assuredly has an effect on their ability to digest food.

The literature abounds with examples of humans ingesting various items to alter the ecology of their guts to enhance digestion, rid themselves of pathogens, or unknowingly to stimulate their immune systems. Although many of these practices are unproven, or unsafe, and so should be avoided, there have been some surprising initial discoveries in recent years that may eventually upend our ideas about what we ingest for good health. For example, although eating clay or dirt is considered an eating disorder in developed countries, its utility in altering gut diversity in the human stomach, at least in some cases, has been shown. And although it might seem silly to recommend giving patients certain elements of fecal matter, or a slurry of particular microbes, these strategies are thought to adjust the overall microbiomes of humans and so have become a major focus of immunological research.

CHAPTER 6

What Is "Healthy"?

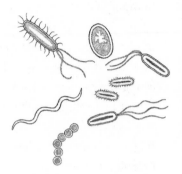

In 2013, the American Medical Association classified obesity as a disease. Obesity is an enigma, because although it is widespread and preventable, it is the single most deadly disorder we know. In many cultures, being overweight is considered a quality of the well-to-do and healthy. Being fat means that you are healthy and wealthy enough to acquire ample food. In fact, in some developing countries stricken by AIDS, overweight sex partners are preferred to thinner ones. In these countries, AIDS is considered mostly a "thinning disease," so overweight sex partners are considered more likely to be unafflicted.

Another example of how definitions of "healthy" are culturally dependent is bilharzia (or schistosomiasis), a debilitating disease in tropical countries that in some countries is the second most socioeconomically devastating illness after malaria. Bilharzia is caused by transmission via freshwater snails of small worms called trematodes. These parasites invade the internal organs such as the bladder, which if damaged by the disease will leak blood into the urine. Bilharzia is so prevalent in some countries that it is accepted as a fact of life. In Egypt, men who are affected don't look at this bloody urine symptom as a problem; instead the bloody urine is called male menstruation and waved off as such. Similarly, in countries afflicted with high infection rates of malaria, this disease is looked on as the norm, because so many people have to live with it.

Clearly our attitudes toward who is sick and who isn't are highly influenced by cultural norms. Not surprisingly, then, this cultural context is also important in discussing how we cope with the thousands of microbes that inhabit our bodies. Indeed, our ideas about sickness or wellness are intertwined with our microbiota, our cultural attitudes, our genomes, and our overall view of the world.

Coevolution of Pathogenic Microbes

What is bad for us and what is not is a tricky question, one that our immune systems have to contend with every moment. There are many cases of pathogens posing strange challenges to hosts and yet coexisting with them. How have some pathogenic microbes coevolved to persist in human populations? As a first step toward answering this question, consider the parasitic microbes that cause malaria in humans. Malaria, a debilitating disease responsible for as many as a million deaths each year, is caused by blood parasites in the genus *Plasmodium*. These single-celled, parasitic eukaryotes live inside blood cells after infecting the human body. Mosquitoes are the vectors for their introduction into our bloodstream.

Because the parasites rely on closely related mosquitoes in a specific genus as vectors, the disease is prevalent in areas where those mosquitoes prefer to live. This reliance on the mosquito vector restricts, for the most part, the range of the parasite to warm wet areas of the planet, such as Africa, India, and Southeast Asia, but it also used to occur in places such as Greece and Italy. Oddly enough, the parasite has more than likely cohabited with us for the entire existence of our species, as well as with many of our common ancestors with other hominids and apes. Even birds and lizards have species of malarial parasites, and although it appears that some parasite species have jumped hosts, it is clear that *Plasmodium* parasites associated with a wide range of vertebrates millions of years ago. Because the preferred place of residence is in red blood cells, the biology of these cells is extremely important to the propagation of the parasite.

Human blood is an interesting and complex collection of cells. The blood cells that are infected by *Plasmodium* are called erythrocytes or red blood cells, and they are the most abundant kind of cell in the blood. Their main purpose is to transport and deliver oxygen to different parts of the body. By examining the genomes of more than 100,000 people, John Chambers and his colleagues showed that there are seventy-five genes that make proteins in the red blood cell. When altered, these proteins can cause various blood disorders in humans, most commonly anemia. One of the most important of these proteins, hemoglobin, is actually a complex of four proteins linked together. Two of the proteins are called alpha-globin and the other two are called beta-globin. Hemoglobin links to iron that in turn binds oxygen, and it is found in very high concentration in red blood cells. The concentration of hemoglobin in a single red blood cell is so high that these cells have shed their nucleus and mitochondria to maximize space in the cell for hemoglobin. The red blood cells have a discoid, lozengelike shape for maximal ability to traverse the circulatory system and they are quite elastic, so they can fit through narrow capillaries.

Mutations in the alpha- and beta-globin genes can lead to changes in the shape and function of the hemoglobin protein. One mutation that causes a single amino acid change in a critical part of the beta-globin protein is involved in a debilitating blood disease called sickle cell anemia. This mutation causes the beta-globin to fold in a detrimental way, leading to a sickle-shaped red blood cell

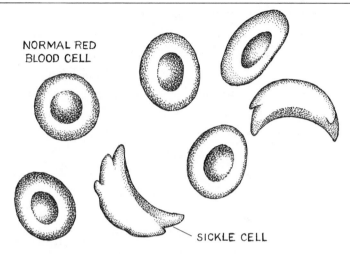

NORMAL RED
BLOOD CELL

SICKLE CELL

Figure 6.1. Normal and sickle-shaped red blood cells. People with a certain mutation in beta-globin will develop the disease sickle cell anemia, which is characterized by sickle-shaped red blood cells.

(hence the disease's name). These deformed red blood cells also are extremely rigid and unable to squeeze through tight spots; instead they get stuck in blood vessels and can stop them up completely (Figure 6.1). Although it takes a normal red blood cell about sixty seconds to make a circuit through the body, the sickled red blood cell is much slower. And it isn't just that the sickled red blood cells get stuck. They also live only about a tenth as long as unaffected red blood cells. Anemia ensues, because the bone marrow that makes new red blood cells simply can't replace the dying cells fast enough.

Human genetics determines to a great extent how a person's hemoglobin will behave. Humans have two copies of each of the twenty thousand or so genes in their genomes (except for males with respect to the genes on their Y chromosome). They inherit one set of copies from their mother and one from their father. So each human has two chances to acquire healthy proteins from their genetic code. If a person has two normal beta-globin genes, then that person's hemoglobin will be normal, and the red blood cells will be lozengelike and highly elastic. But if a person has two hemoglobin S genes, that person's hemoglobin will sickle, and he or she will show extreme sickle cell anemia. In most places on the planet, the hemoglobin S allele is so detrimental that it is eliminated from populations because of its devastating effects on the fitness of individuals with two copies of it. But in populations where malaria exists, this allele persists and for a good reason.

It turns out that if a person has one copy of the hemoglobin S allele and one copy of the normal allele, those red blood cells that do not have enough oxygen will sickle, whereas those that have enough oxygen will stay a normal shape. Further, when the *Plasmodium* parasite is living in red blood cells, it munches up the hemoglobin in the cells, thereby starving them of oxygen. This means that a *Plasmodium* infection in a person with one hemoglobin S allele and one normal allele will lead many of the infected red blood cells to sickle, and it is thought that these sickled cells are more quickly destroyed by the immune system. The parasites are thus removed along with them before they can replicate and complete their life cycle. Humans with both one hemoglobin S allele and one normal allele are thus more resistant to malaria. Where malaria is prevalent, then, the individuals who are heterozygous are the most fit in the population, and the population will evolve to maintain both the hemoglobin S allele and the normal allele for as long as the malaria parasite sticks around.

The case of sickle cell anemia and malaria is not the only one where a pathogen has influenced the genetics of our species. Trypanosomes are another kind of nasty single-celled organism that can infect humans. These parasites cause debilitating and often fatal diseases such as sleeping sickness and Chagas disease. Like malarial parasites, they have complex life cycles and are transmitted through insects. When a sleeping sickness–causing trypanosome infects a human, it enters the bloodstream through the bite of a tsetse fly and first causes fever, headaches, pruritus (reflexive itchiness), and swollen lymph nodes—indicating the parasite's presence in the bloodstream and lymphatic system. Stage two of the disease is characterized by sleep disorders (from which the disease gets its name) and a number of neurological disorders such as tremors, motor weakness, and symptoms similar to those of Parkinson's disease—which demonstrate that the parasite has made it across the blood-brain barrier and is wreaking havoc in the brain. The human genome has evolved to combat this parasite. But the story doesn't end there: humans who have evolved to more effectively fight off trypanosome infections are also generally more susceptible to kidney disease.

To tease out this interplay of genetics and environmental factors, we must first report that medical scientists, including Giulio Genovese and his colleagues, have shown that two forms of kidney disease in African Americans can be traced to mutations in a single gene of the human genome. These two disorders—hypertension-attributed end-stage kidney disease (H-ESKD) and focal segmental glomerulosclerosis (FSGS)—are associated with specific mutations in a gene called apolipoprotein 1, or APOL1. This protein is a minor player in making up high-density lipoprotein (HDL) or what is also known as "good" cholesterol. The two mutations' frequent occurrence among African Americans, however, makes little sense until one looks at what else the apolipoprotein gene does. Although it is part of HDL, it is also known as a serum factor that can lyse trypanosomes. Although research so far has not indicated that the APOL1 alleles are under heterosis, like the sickle cell alleles, it is a good bet that these two APOL1 variants have hung around in African populations as the result of some selective advantage for eliminating trypanosomes from the blood system.

There are at least twenty human genetic disorders that researchers have suggested are linked to pathogen resistance and probably many more to come. Although some researchers don't buy the heterozygote advantage argument, as we learn more about the human genome and about the biology of infectious microbes and how they interact with us, the evolutionary perspective only becomes more salient.

Imagine you are a benevolent alien living seventy thousand years ago, cruising around in your spaceship, and you find Earth. You land, and you find a strange upright hairless species with a pretty advanced way of verbally communicating with each other. You realize that this species is restricted to warm, relatively wet areas of Africa in small isolated populations and that they are prone to many genetic disorders. The genetic disorders have gone to high frequencies in these small populations. You surmise this because of the small population size of this strange species and because rare alleles have by chance gone to high frequency in the small populations in a process known as genetic drift. There are other populations of upright hairless humans in other areas of the planet, but they do not have the advanced language skills that the populations in Africa have, nor do they have the same genetic disorders. Not having a Star Trek prime directive (in other words, you *are* allowed to interfere with the development of the alien civilizations you are encountering), you decide to see if you can help the health and welfare of this strange species. You somehow obtain DNA samples from all of the populations of humans in the African area where you have landed. You discover that there are two major diseases that are affecting the strange species. One is a blood anemia and the other is a killer kidney disease. You pinpoint the genetic lesions responsible for the two diseases and, being the benevolent alien species you are, you decide to genetically engineer the alleles for the two diseases out of the small populations.

You do your work, and then speed off to another galaxy to explore. Taking relativity and time travel into account you decide to return to Earth seventy thousand years later. You land and are entirely shocked to see that the upright hairless species you tried to help has gone extinct.

Marty Kreitman, a biologist at the University of Chicago, first proposed this scenario and pointed out that you, the good-hearted alien, by eliminating the blood disorder and the kidney disorder, also eliminated the genes for resistance to malaria and trypanosomiasis from the struggling species. The species you tried to help did not survive the double whammy of the *Plasmodium* and trypanosome infections it was experiencing. Although this is a science fiction story, it is pretty close to reality. The interactions of our genes, our microbes, and their interactions (their ecology) are tightly interwoven.

Helicobacter pylori

Getting old affects many parts of the body. Perhaps one of the worst effects of advancing age, at least to a gourmet, is the one on digestion. Even if as youngsters we could plow through whole pizzas smothered in pepperoni, onions, jalapeño peppers, and spicy tomato sauce, the older we get the more likely it is we will choose instead the regular cheese pizza with mild tomato sauce to avoid the late-night heartburn. The lifestyles of bacteria and how we have treated them cause many of us to change our diets as we get older. The saga of one species in particular, *Helicobacter pylori*, illuminates well the tight coevolution we have with microbes (Figure 6.2).

We have been living in relative harmony with *H. pylori* for at least 100,000 years. At that early time in our evolutionary history, humans were still in Africa. (Migration of our species out of Africa would not occur for another 40,000 to 50,000 years.) There is pretty good evidence that *H. pylori* was present in our ancestors this far back and might even have been an advantageous inhabitant of our

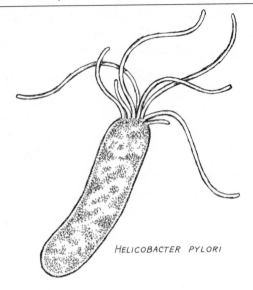

HELICOBACTER PYLORI

Figure 6.2. *Helicobacter pylori.*

guts. Martin Blaser at New York University School of Medicine has suggested that *H. pylori* might have protected our ancestors from diarrheal diseases and, more important, might have been essential for the more efficient maintenance of everyday bodily functions. Humans living at this time more than likely relied on *H. pylori* as part of their gut microbiota to maintain healthy lives. About ten thousand years ago, the lifestyles of humans changed drastically from hunter-gatherer strategies to more domesticated forms of sustenance and a more centralized population structure. Along with these changes came changes in diet and probably changes in gut microbiota. During this period of change in the way humans lived, *H. pylori* strengthened its coevolutionary hold on humans. One thousand years ago, humans were increasingly moving to cities, and close contact between humans became a normal part of everyday life. Until this point in time, probably every adult human had

H. pylori in his or her gut. In fact, if these humans were anything similar to us, *H. pylori* would have been one of the major inhabitants of their gut microbiomes.

Much more recently, about a hundred years ago, humans experienced a revolution in how they viewed and maintained their health, because of improved hygiene practices and the advent of antibiotics. These changes resulted in a systematic elimination of *H. pylori* from the gut microbiome. Today only about 10 percent of children have any *H. pylori* in their gut microbiomes. Given the ubiquity of *H. pylori* in the stomachs of humans before about a hundred years ago, this is one of the most extreme changes in microbial community ecology on record. There were certainly costs to harboring *H. pylori,* including a higher risk for peptic ulcer disease and gastric cancer. These diseases occur at a greater rate when *H. pylori* is causing inflammation of the stomach wall and affecting the amount and distribution of gastric hormones, changes that in turn lead to drastic shifts in one's digestive physiology.

Helicobacter pylori, however, isn't a total villain. Its presence decreases the risk for GERD or heartburn caused by gastroesophageal reflux disease, as well as a slew of other diseases (GERD is also linked to asthma and esophageal adenocarcinoma, for example). So by getting rid of *H. pylori* because it causes stomach ulcers, we increase the risk for these other ailments. *Helicobacter pylori* also influences the production of gastric hormones, which are essential in the metabolism of leptin and ghrelin—small peptides that are involved in enhancing our sense of hunger and satiety and hence are involved in the biology of appetite. In fact a lack of leptin and ghrelin is associated with obesity and type II diabetes. Clearly what was considered a pathogen in the past century is now being seen as part of a more complex system.

Our definition of what a pathogen is clearly needs to change as we become aware of the many complex interactions that microbes have with our bodies. To accomplish this goal, we should consider the ecological interactions that communities of microbes have with us, on and in our bodies, and in areas around us. We will focus on three parts of the human body in this respect—our guts, where the majority of microbes living in our bodies exist; the genitalia, where human behavior has a huge influence on microbial ecology; and, perhaps surprisingly, our brains, where microbial ecology has been shown to have a distinct effect on our behaviors. Because what we eat is one of the most important and influential changers of the microbial ecology in our bodies, it is no wonder that the microbiome of the digestive tract and the ecological parameters that influence it have been a major focus of researchers working on microbiomes and human health. The microbial makeup and ecology of the digestive tract have been implicated in everything from cancer to obesity to our behaviors.

The Ecological Perspective

Ecologists have a long and storied tradition in biology, and their contributions to our understanding of the interactions of organisms have been substantive. Ecologists work at the level of interactions both within communities and between communities. One of the basic goals of ecological community analysis is to describe different levels of variability in communities. A first step in this way of studying things is to characterize the species that belong to a community. For microbial communities associated with the human body it should be easy to see that, in the past, community-level studies of microbes were incomplete, if only because until a few years ago characterization of the species in microbial communities on and in the human body was restricted to species that could be cultured.

In Chapter 2, we considered the diversity of communities in two different ways—alpha diversity and beta diversity. Alpha diversity refers to the diversity within a given community. For instance, if the belly button is a good habitat for lots of different microbes, then there will be a high alpha diversity in belly buttons. Beta diversity refers to the degree of species variation between described communities. Thus there is beta diversity between belly buttons of different people, but there is also beta diversity between a belly button community and, for example, an armpit community on the same body or on different bodies.

When comparing microbiomes it is fascinating to realize that the beta diversity for communities on a single human body (for example, the belly button versus the armpit) exceeds the beta diversity of the same kind of communities on different human bodies. It is important, then, to keep these two kinds of beta diversities distinct in our minds. Although we have mentioned core microbiomes elsewhere, we need to realize that if the concept of a core microbiome is to be useful at all, it needs to be defined on the basis of the beta diversity that exists either between specific habitats on and in the human body or between different individuals for the same habitat.

Another important factor in understanding how ecology is important in studying the microbiome comes from the classic characterization of ecological communities of animals and plants over time. Change in habitat conditions such as pH or temperature or invasion of a habitat are a few of the many possible ways that the environment of a community can be altered, and communities respond to such changes in three major ways. The first is by developing what are called resistant communities. These communities are relatively unaffected by environmental changes, and their alpha diversity hovers around a well-characterized average diversity. The second is by creating resilient communities, which fluctuate wildly with respect to their overall alpha diversity upon being disturbed, but return to the initial alpha diversity in due time. These communities are considered somewhat stable, even though they might go through some drastic fluctuations in species composition. And the last kind of community response is to go with the flow of the disturbance and alter the species composition for the long term, creating a distinctive new alpha diversity. This kind of an ecology or community dynamic occurs as the result of what microbial ecologists call dysbiosis. Whereas symbiosis involves species interacting in harmony to the point of helping each other, dysbiosis occurs when a normally symbiotic set of species or communities are entirely disrupted and the microbiome profile changes to a novel and unusual mix of microbes. Dysbiosis leads to unexpected interactions, and if these interactions occur in or on the human body, they can sometimes be pathogenic.

Dysbiosis is complex, and there is a wide range of ways that it can happen to a healthy ecosystem, whether that ecosystem is a rainforest or your gut. The presence of intruders or interlopers, or the disappearance of any integrated member of the healthy microbial community, can change the community makeup; a good example is when the microbes that cause bacterial vaginosis disrupt the vaginal microbiome. Second, shifts in the microbial "balance of power"—that is, in the relative species abundances of integrated members—can also disrupt a working microbiome. This kind of dysbiosis is common with respect to our guts when shifts in diet occur or when severe infections occur and antibiotics are subsequently applied to treat infections. The horizontal transfer of genes causes the third kind of dysbiosis, in which instead of a change in species composition, a shift occurs in the kinds of genes that are present. This third process has been implicated in some of the more pathogenically interesting cases of microbial infection.

Bacterial Blooms and Sickness

One sure sign that an aquatic or marine ecosystem is out of whack is a bloom. Probably many have seen those fantastic pictures from

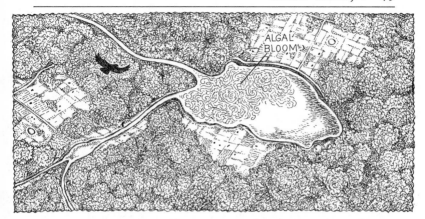

Figure 6.3. Drawing of a pond with a severe algal bloom.

satellites of algal blooms covering vast areas of the ocean or of lakes (Figure 6.3). Our guts and other inner areas of our bodies are also prone to blooms, but not the algal kind. In the human gut a bloom happens when species that are normally present in low numbers grow to large numbers. A low-density microbe in the gut is one that is represented there by very few cells. So for instance, if you took a milliliter of goop from your stomach and plated it out on agar and fewer than a hundred or so microbes of a particular species grew up, this would indicate that that species was a low-density species. By contrast, common species in the gut will be present in the millions in one milliliter of stomach goop.

What kind of environmental changes will make the gut ripe for a bloom? For a pond or an area of the ocean it might be a sudden change in temperature, an abrupt shift in pH, an infusion of nutrients, or the introduction of a pollutant that then eliminates the usual cohabitants of the pond or ocean ecosystem, leaving the ecosystem open for troublemakers.

Microbial blooms in the gut are also caused by environmental changes, but in these cases, the changes can be phenomena such as drastic changes in diet. It's also known that the genetic predisposition of the human owner of the gut can contribute to blooms. Microbial blooms in the gut are particularly nasty, because they produce a cascade of negative effects. If any of the microbes in the bloom are resistant to the defenses that our guts throw at them, then the resistant microbes will probably be quite strong and natural selection will favor their proliferation. By the same logic, if the bloom contains several kinds of resistant bacteria and there are fitness differences among these nasty microbes, the one that is most fit, and possibly the most pathogenic, will proliferate, causing a cycle of dysbiosis that leads to sickness. Just to demonstrate the potential for more ecological disaster, imagine that the microbes that are successful can horizontally transfer genes that confer the fitness effect. In addition, imagine being treated with an antimicrobial that wipes out much of your microbial ecosystem, but not the successful, unwelcome microbes. Both scenarios will result in a population of microbes that are raring and ready to cause infectious disease.

Bubble Mice

Researchers have bred some very interesting strains of mice to observe how microbial blooms take over in the guts of these rodents. The mice are like living Petri dishes in which microbiologists do experiments. Mice have been bred to have abnormal immune systems or to be missing certain proteins and they are then tested to see how they, and their gut ecologies, respond to all kinds of perturbations. The ultimate way to look at how microbiome ecologies change in the living gut, however, is to use animals that are blank slates with respect to microbes: "bubble mice."

The first measure any self-respecting microbiologist takes before conducting an experiment is to sterilize all of the materials, glassware, Petri dishes, and the general area where an experiment will be run. Standard procedures include autoclaving all tools at temperatures over 250° C. Researchers working with mice or other experimental animals obviously cannot use this approach to get an experimental subject that is germfree. So they have resorted to raising mice in sterile bubbles much like the "bubble boy" whom Jake Gyllenhaal played in the movie of the same name. How are the mice delivered from their mothers without becoming contaminated with microbes? Vaginal delivery of an infant, no matter what species, introduces a whole new community of microbes, so the mice are born via Caesarian section under highly sterile conditions. The baby mice are then nursed by surrogate sterile mothers under sterile conditions and placed in sterile living quarters that resemble the home of the cinematic bubble boy. The researchers who use them affectionately call them gnotobiotic mice ("gnoto" from the Greek "gnosis," or "known"), but we prefer "bubble mice."

There is also a category of carefully bred mice called "pseudo-germfree." These mice are given broad-spectrum antibiotics containing a cocktail of antimicrobials that deliver a double, triple, and even a quadruple whammy to their intestines. One of the cocktails that includes ampicillin and neomycin can eliminate 90 percent of the microbes in the gut of a mouse, but the quadruple cocktail of vancomycin, neomycin, metronidazole, and ampicillin can get rid of almost everything in the gut. Once these mice have been treated with the antibiotics, they are raised under highly controlled and sterile conditions.

Because researchers use highly inbred strains of mice for these experiments, the genomes of the normally raised and germfree-raised mice are essentially identical. If a researcher wants to see how a particular mutation affects colonization of the gut by microbes, he or she simply constructs a mouse strain with the mutation and raises one group of mice from the strain in the bubble and another under normal conditions. It is then possible to deliver microbes quite precisely to the bubble mice to see how certain microbial interactions affect the growth, development, and health of the mouse. Bubble mice are not without their initial problems and differences from normally raised mice, however. For instance, germfree mice have a bigger cecum (part of the large intestine) than normally raised mice, and not surprisingly, bubble mice have very underdeveloped and understimulated immune systems. They also do not defecate as well as normally raised mice, and female reproduction cycles are abnormal. (Significantly, if a normal microbiome is introduced into the mice, all these symptoms quickly disappear.) The initial physical and physiological problems suffered by the germfree mice therefore have to be considered when running experiments with them.

As mentioned earlier, the microbiomes of the stomach, intestines, and rectum vary both in their microbial species composition and in the number of bacteria present. We will focus here on the intestine because it has the biggest microbiome, in terms of sheer numbers of bacterial cells, of all the human digestive tract habitats. A properly functioning intestine has to achieve two goals. First, it must ensure that there are no interlopers that can interfere with the smooth running of the digestive process and the passage of material to the large intestines and eventually out of the body. To do so, it must have mechanisms for eliminating dangerous infectious microbes, while simultaneously retaining the microbes that it uses for digestion. Second, and more important, it needs to absorb nutrients from the food and liquid items that have passed through the stomach. The human intestine is very well suited for both of these tasks. It is incredibly long and has a huge interior surface area for absorption, nearly two hundred square meters. In addition, lining the entire intestinal tract is a layer of mucus, which as we learned in Chapter 4 is important for the health of the beneficial microbes in this part of the digestive system (Figure 6.4). The mucus performs several jobs, such as lubricating the passageways so that digested food can pass

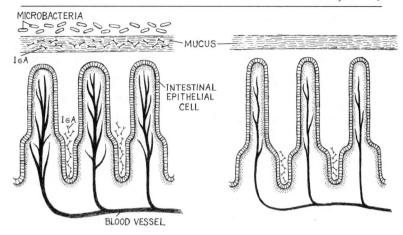

Figure 6.4. The cellular structures of the lining of the human stomach.

through more easily, and it harbors the mucosal immune cells that defend the intestine from dangerous invaders. The mucous layer of the small intestine is very thin, only about a hundredth the thickness of the mucous layer of the colon. It is comprised of a group of proteins appropriately called mucins, of which mucin2 (MUC2) is the major player. These proteins are glycosylated, which simply means they are covered with glycan molecules. These molecules are a lot like sugar molecules; hence the mucin proteins in mucus can be thought of as being "sugar coated." The sugar-like coating and large size of mucins means that they behave a lot like jelly, which is helpful for lubricating the passage of digested material. The mucous layer itself is partitioned in two layers. Because of the way the mucins in the mucous layer interconnect in a kind of lattice, microbes are restricted to the area above the interconnected mucins. The localization of the microbes in this upper layer is important and means that the habitat and ecological characteristics attract specific kinds of microbes.

As J. Travis puts it, the innate immune system is like a "bouncer at a nightclub, trained to allow the right microbes in and kick the less desirable ones out." The mucous layer of a healthy small intestine is indeed like the ultimate bouncer at a nightclub: it discriminates between microbes that are there to help and those that are there to harm, and it ensures that there are no blooms—in other words, that the area is not overcrowded. To regulate the populations of microbes in these ways, it floods itself with immunoglobulins (IgA in particular) and small antimicrobial proteins, both of which are deadly to some microbes.

Unhealthy intestines are a different story. Experiments with bubble mice show that their microbe-free intestinal mucous layers have very specific differences from those of typical mice. The mucous layer is considerably thinner than in normally raised mice. In addition, the fingerlike protrusions of the inside of the intestine called microvilli are narrower and longer in sterile mice than in regular mice. And because sterile mice have a relatively unstimulated immune system, products of the immune system, such as IgA molecules and antimicrobial peptides, if they are present at all, will be in lower concentrations than those products in normally raised mice.

Two interesting manipulations of normal and sterile mice tell us a lot about how the system works. If one simply introduces into a sterile mouse molecules derived from bacteria such as peptidoglycan or lipopolysaccharide, the molecules alone will stimulate the immune system of the bubble mouse, and most of the problems we have listed will be rectified. Lipopolysaccharides are simple molecules that are found in abundance in the membranes of certain kinds of bacteria that live in the gut. When bacteria get into the gut, some of them are digested, but lipopolysaccharides from their destroyed membranes float into the bloodstream. When lipopolysaccharides enter the bloodstream, the host has an immune response that causes inflammation. Some species have very nasty lipopolysaccharides that do much more damage than those of other species.

For instance, microbes in the family Enterobacteriaceae, such as *E. coli,* have lipopolysaccharides that are a thousand times more toxic than those of bacteria from the family Bacteroideaceae, which are themselves prolific producers of lipopolysaccharides in our guts. In fact, obese mice have huge amounts of bacterial lipopolysaccharides in their blood plasma relative to lean mice. The obese mice also show intestinal inflammation as a result of the elevated levels of lipopolysaccharides. In fact, if one isolates the lipopolysaccharides from *E. coli* and injects them for just four weeks in normal mice that are fed normal mouse chow, they become obese, inflamed, and nonresponsive to insulin. If these animals are then treated with antibiotics, their gut microbiomes will be altered, the lipopolysaccharide concentration goes down, and they do not develop obesity or insulin insensitivity. These data demonstrate that the immune response to these simple molecules that come from microbes in the gut are enough to stimulate the immune system to mature and to tweak the defense system of an individual. Apparently, just the "smell" of bacteria is enough to activate the system.

In other experiments, the mucous layer itself is removed in normally raised mice to see what its role is. Making a mouse strain with a silent MUC2 gene (the gene that makes the protein that is the major component of the mucous layer) is the experimental approach. Such strains of mice where a gene is rendered nonfunctional are called knockouts, and when the MUC2 knockouts are exposed to microbes, the intestinal wall becomes overgrown with microbes. These mice are highly susceptible to colitis, an inflammation of the colon. The two experiments together indicate the importance of the mucous membrane in regulating the proper function of the intestinal wall. The microbes typically present in the gut microbiome are essential for the normal maturation of the gut cells, and for preparing the gut cells to be efficient "bouncers."

Obesity, Diet, and Genetics

Obesity is a tricky disease to cure. Although it has a great deal to do with diet, most people find it difficult to regulate or modify their intake of food, so other approaches are being tried. One of these approaches involves the microbiota involved in obesity and the genomics of obesity. Germfree mice, whether gnotobiotic or pseudogermfree, have been used in this focus. But one of the seminal studies on how diet influences gut microbiota and obesity was accomplished without germfree mice. Researchers fed mice a consistently high-fat diet, much like Morgan Spurlock did for his *Supersize Me* documentary film. Like Spurlock, the mice did not fare well physically. Although Spurlock tried subsisting on food from a single fast-food corporation for only thirty days, the mice in this experiment were fed a high-fat diet for eighty-four days straight. After the twelve weeks of the supersize-me diet, the mice were obese (with a 60 percent increase in adiposity or fat) and resistant to insulin. Being insulin resistant is particularly dangerous to mouse (and human) health, because it leads to diabetes.

Spurlock, who was thirty-two years old at the time of the experiment, gained twenty-four pounds, experienced psychological dysfunction (depression and loss of interest in sex), massively altered his liver chemistry, and ended up with a cholesterol count that was extremely high. It took Spurlock no fewer than fourteen months to return to his former weight, cholesterol count, and sex drive. Spurlock's experiment on himself occurred in 2003, at about the same time that a high-throughput examination of microbiomes was being envisioned. We are pretty sure that he wouldn't want to repeat the experiment and have gut microbial samples taken along the way, but it would have been really interesting to have followed his microbiome through his supersizing. Not to worry, though, because the scientists who did the mouse experiment did collect gut and fecal material to follow the change in microbiomes on the path to obesity and, as it turns out, on the path back to normal weight (when the obese mice were placed on a strict diet of normal mouse chow for ten weeks). Ultimately the mice with the forced diet changes lost weight, regained their ability to respond to insulin, and were, in general, healthier than when they had been obese. A control cohort of mice was fed normal chow throughout the experiment, and because they were genetically identical to the mice with the forced diets, they posed a perfect baseline for a comparison of the microbiomes (Figure 6.5).

The results of changing the diet of the obese mice were expected: that is, the kinds of bacteria in the supersized mice quickly reached a distribution that was very different from the control mice. But after four weeks of being forced to return to only normal chow following their supersizing, the species composition of the gut microbiomes of the forced diet mice started to converge on that of the control

Figure 6.5. A typical wild-type mouse (*left*) and a typical obese mouse (*right*).

mice—and by ten weeks later, the microbiomes of the control mice and the forced diet mice were indistinguishable. This study showed for the first time that the microbiomes of the guts of obese individuals were significantly different from nonobese individuals and that a return to a healthy weight was accompanied by a return to a specific microbiome.

This revelation led to many experiments to pin down the exact nature of the interaction of obesity with diet and genetics. In pursuit of an answer to this question, Peter J. Turnbaugh and his colleagues at Washington University in St. Louis took genetically identical sterile mice and divided them into two cohorts. One cohort was fed microbiota from the guts of obese mice, and the other cohort was fed microbiota from the guts of lean mice. Both cohorts received the same kind of chow, so their nutrition differed only in the microbes they were being fed. After a sustained diet of microbes and chow in this experiment, the two cohorts were analyzed, and the obese-microbiome-fed mice showed a much higher adiposity than the lean-microbiome-fed mice. As discussed in Chapter 4, it is well known that obesity is the outcome of the relative abundance of the two dominant bacterial divisions, the Bacteriodetes (B) and the Firmicutes (F). Obese mice have a lower ratio of B to F microbes in the gut microbiome. The sterile mice that had been fed obese microbiome had a lower B to F ratio than the sterile mice fed the lean microbiome. In addition, Turnbaugh and his colleagues determined that the obese microbiome obtains energy from food more efficiently than the lean microbiome does. Obesity itself then arises from this capacity of the microbiome for higher energy processing.

Although the B to F ratio tells us a lot about whether an individual will be obese or lean, it doesn't entirely explain which microbes are making a difference. In 2013, researchers showed that to change the ratio of B to F is also to change the numbers of a normally commensal beneficial microbe that lives in the mucous layer called *Ak-*

kermansia muciniphila. Specifically, the number of *A. muciniphila* cells decreases in the guts of obese individuals in both mice and humans. This phenomenon is known, because mice fed high-fat diets have a low *A. muciniphila* and higher lipolysaccharide concentration. Well, why not simply feed large amounts of *A. muciniphila* to the obese individuals? Using some of Koch's postulates, researchers tried exactly just this experiment. They fed mice reared on high-fat diets live *A. muciniphila* for four weeks. They also fed the same kind of mice heat-killed *A. muciniphila* for four weeks. The mice fed live *A. muciniphila* showed lowered lipopolysaccharides, lowered adiposity, and lower glucose in their plasma (which is related to the responsiveness of cells to insulin). These lowered levels were pretty much the same as those for a control population fed normal chow. The mice fed heat-killed bacteria, by contrast, showed no reduction in adiposity, lipopolysaccharides, or insulin responsiveness. Upon further examination of the mice fed live *A. muciniphila,* researchers found that genes involved in the breakdown of lipids and increased adipose tissue metabolism (in other words, genes involved in getting rid of fat) were expressed at a higher rate than in control mice.

High-fat diets are notorious for disturbing the mucous membrane of the intestine. In fact, they partially destroy it and the underlying tissue of the intestine itself. Apparently *A. muciniphila* restores the mucous membrane that is badly damaged by a high-fat diet. By examining the intestinal lining of mice fed *A. muciniphila,* researchers discovered that the microbe actually restores the intestinal mucous layer and reinforces the barrier between digested food going through the intestine and the intestinal tissue itself, leading to a reversal in the pathology induced by obesity and diabetes. To shore up this idea, researchers have fed to mice high-fat diets supplemented with a molecule called oligofructose. This molecule is used in other therapeutic contexts as a probiotic that can help regulate gut microbes. Remarkably, the numbers of *A. muciniphila,* even in these mice with high-fat diets, increases substantially. Whether these effects will also occur in humans is still to be determined.

Bubble mice have also been used to study the impact of sugary or "Western" diets. When germfree mice are fed a sugary Western diet, they avoid obesity. In fact, sustained feeding of sterile mice on the Western diet did not result in significant weight gain compared with controls fed the boring normal chow. Since the experimental mice are sterile and were fed sterile food, no microbes were present in the guts of these animals to influence the processing of energy from the food. Thus the microbiome (or in this case, the lack of one) is influential in determining whether an individual will be obese or not. A variation on this study using a cohort of sterile control mice fed normal chow and a cohort of sterile mice fed the sugary Western high-fat diet for eight weeks revealed that the high-fat Western diet produced mice that had significantly higher adiposity. Although this result at first might seem contradictory to the results of the study where obese and lean microbiomes were fed to mice, it turns out that after the first eight weeks the mice fed the Western diet started to lose weight and after a second eight-week period were not significantly fatter than the control mice that were fed on the normal chow.

Although these studies utilized mice, they do enlighten our perception of how our gut microbiomes might be reacting to diet and, more important, how our gut microbiomes might be responsible for the development of obesity.

What about genes? The involvement of genes in obesity is well known, and some of the experiments just described actually address this question. The obese mice from which obese microbiomes were extracted to feed to other mice are what are called leptin-deficient. They have mutations that knock out the function of leptin, which along with grehlin, are two proteins that are important in obesity. The immune system of a human or mouse host is intricately involved in the regulation of microbes in the gut and, of course, genes ultimately code for our immune systems. Leptin itself influences the immune response of the gut and is in fact a cytokine that induces and controls some of the acquired immune response by regulating what kinds of T cells are produced in the thymus. It also affects neutrophil production and migration. This means that if a mouse is leptin-deficient or has been knocked out for leptin, the lowered leptin concentration more than likely will influence the activity of the immune system.

Other immune system effects of diet and genes have been observed in both mice and humans. Small molecules called short-chain fatty acids from bacteria have a huge influence on the immune response. We cannot digest a lot of our diet with our own gene products. For instance, we simply do not have the genes to make the proteins or enzymes to break down plant polysaccharides. But the microbes in our gut microbiome do—through the fermentation process. Fermentation is one of the major pathways that microbes undertake to provide themselves nutrition. The endproducts of bacterial fermentation are short-chain fatty acids, so their presence is a giveaway that certain microbes are present. Our immune systems have learned to detect the short-chain fatty acids and react with a number of innate immune system responses. These immune system responses in turn influence the composition of the microbiome.

The Vaginal Microbiome

Our earlier census of microbes in the vaginas of human females ended with the observation that there are basically five major kinds of healthy vaginal microbiomes (see Chapter 4). Or are there? Whether a vaginal community is considered healthy or not is usually based on its predisposition for a sexually transmitted infection. But this criterion turns out not to be a good one. The microbiota of the vagina therefore has become an example of why we need to redefine pathology in the age of the microbiome.

The most common pathology of the human vagina is a disorder known as bacterial vaginosis. Although bacterial vaginosis is not the most devastating disorder a woman can have, it is thought to be a risk factor for sexually transmitted infections. Most cases of bacterial vaginosis are cleared either by oral antibiotics or by the application of antibacterial creams. The prevalence of bacterial vaginosis in the United States is nearly 30 percent, meaning that 30 percent of women have had it at least one time in their lives. In addition to being a risk factor for sexually transmitted infections, bacterial vaginosis has also been correlated to the transmission of HIV and pelvic inflammation, among other more severe pathologies, so it is a matter of some concern. Before the vaginal microbiome was characterized, clinicians would determine the risk for and presence of bacterial vaginosis based on two kinds of measurements, the Amsel and Nugent scores, which basically measure the quantity of *Lactobacillus* in the vaginal cavity. Remember that *Lactobacillus* species in the vaginal cavity are actually beneficial, because they ferment nutrients and produce lactic acid, which lowers the pH of the vagina and restricts what microbes can grow there. The standard thinking has been that the presence of *Lactobacillus* sort of fine-tunes the ecology of the vagina and limits the growth of other bacteria that can cause bacterial vaginosis, yeast infections, and sexually transmitted infections. Although species from the genus *Lactobacillus* do make up a large part of some vaginal microbiomes, they are by no means the only kinds of microbes present, and, in fact, in some healthy vaginal microbiomes, species in the genus *Lactobacillus* sometimes aren't even the most prevalent microbes in the vaginal community (see Chapter 4).

Clearly it was time for a redefinition of "vaginal disorder." The Amsel and Nugent scores used to characterize the vagina's ecology are based entirely on how abundant microbes in the genus *Lacto-bacillus* are in samples from the vagina. Whereas it is true that all women with lots of *Lactobacillus* in their vaginal microbiomes are spared from bacterial vaginosis, the converse is not true. In other words, many women with low *Lactobacillus* counts (and hence positive Nugent scores, which have been used to indicate a problem) do not manifest the symptoms of bacterial vaginosis. So the healthy vagina cannot be defined on the basis of how abundant microbes in the genus *Lactobacillus* are in the vaginal microbiome. With respect to the microbiome, there is more than one way to have a healthy vagina.

By the same token, then, there is more than one way to have an unhealthy vagina or a vagina that manifests bacterial vaginosis. In other words, bacterial vaginosis, which is considered an unhealthy condition, is what we might call a broad-spectrum disorder. This broad-spectrum idea has become a prevalent one in modern medicine and has become very useful for teasing apart the definition of a disorder and a modern view of what is healthy and what is not. By lumping all bacterial vaginosis cases together as one disorder, clinicians have missed a lot of its nuances, and this oversight has led to a tendency to overdiagnose the occurrence of the disorder in women. Even worse, by lumping all bacterial vaginosis cases together, the ecological complexity of the vaginal microbial community was missed. In the broad-spectrum paradigm, by contrast, bacterial vaginosis, as defined by risk for sexually transmitted infections, is no longer the focus of research, but rather the focus becomes a more comprehensive understanding of the ecosystem that is the vaginal cavity. Detailed ecological analyses of this and other important human body ecosystems are perhaps the only way to understand the complexity of the disorders that are there.

To this end, Bing Ma and colleagues at the University of Maryland and the University of Idaho have been conducting longitudinal studies to characterize healthy vaginal microbial communities in individual women over time. They examined the vaginal community composition in healthy women of reproductive age by sampling thirty-two women twice a week over a sixteen-week period, recording the time and length of menses for each woman, as well as the microbiome composition of each sample. As with the initial microbiome characterization of the vaginal cavity discussed in Chapter 4, there was a significant degree of beta-level diversity among women. In addition, the longitudinal study showed that some subjects experienced dramatic shifts in the composition of their vaginal microbial communities over short periods of time, while other subjects' vaginal microbial communities stayed rather stable. In Figure 6.6, for instance, subject 2 shows very stable vaginal communities over the sixteen-week sampling period. Subject 1, by contrast, shows a radical shift in microbiome composition beginning at around week 8, or halfway through the study. Note that in all of the profiles in the figure there is a spike in species composition changes about every four weeks. These spikes correspond to menses, which was shown by this longitudinal study to be the major factor in creating dysbiosis in the vaginal communities. Periods of the menstrual cycle when there are high levels of estrogen or when high levels of estrogen and progesterone overlap, however, are characterized by the most stable microbial vaginal communities. The study also kept track of each woman's sexual activity, and while this factor caused some shifts in the composition of the microbial communities, the impact of sexual activity was nowhere near that of menses. The dynamic changes that are part of the normal reproductive cycle of women make characterizing the vaginal microbiome daunting, but they also underscore the importance of understanding the bigger picture, as opposed to basing an understanding of a disorder like bacterial vaginosis on a comparison of single points in time, such as when the microbiomes of women with bacterial vaginosis

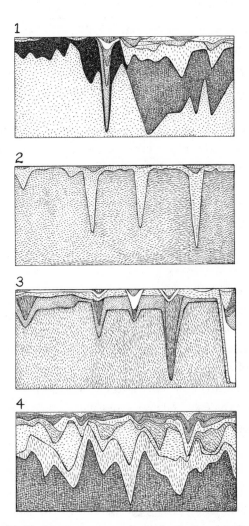

Figure 6.6. Plots showing the turnover of microbes in the vaginal microbial community of four women. The different shades represent different kinds of bacteria; the proportions of those bacteria are plotted on the vertical Y axis; and the day the sample was taken is plotted on the horizontal X axis, which spans the sixteen weeks of the study. The periodicity of the plots (that is, the occurrence of dips of some of the microbes at regular intervals) is due to the women's menstrual cycles, which occurred about every four weeks.

are contrasted with those of women without it. Well-designed longitudinal studies to understand the ecology of specific habitats on and in the human body not only promise to expand our knowledge of day-to-day healthy interactions, but also will give a baseline from which to understand shifts that occur as a result of pathology. The hope is that with this broader and more complete understanding of this complex ecosystem and others associated with the human body, diagnoses of pathologies will become more accurate and novel treatments for such pathologies will be both easier to produce and more efficacious.

The Stomach and Depression

The famous adage that the way to a man's heart is through his stomach is true if the heart is in the head. (It's the same way to a woman's heart.) The connection between the gut and the brain, termed the gut-brain axis, is real and fairly well understood. When someone ingests food or drink, more than just chewing, drinking, and digesting are occurring. Food triggers a myriad of reactions in the mouth, gut, pancreas, and brain. Actually, our reaction to food occurs even before we ingest it, because just the sight or smell of food triggers physiological and hormonal reactions that affect our guts and brains. Microbes have an extremely important influence on the information that is shuttled along the gut-brain axis.

One of the major roles of this gut-brain axis is to regulate appetite. In order to survive, the body must know when it is hungry and when it is full. Hence a complex system has evolved to ensure that the communication along the axis is quick, precise, and efficient. The gut communicates with the brain through hormones that are secreted into the endocrine system mostly from the stomach, but also from other parts of the body. Several of these hormones—including insulin, ghrelin, obstatin, cholecystokinin, and glucagon-like peptide—have been tied to regulating our appetites and in some ways have created a second brain in our bodies: our stomachs. When the brain encounters these hormones, specific responses are induced for seeking or avoiding food. The major neural pathway that allows our stomachs to communicate with the brain is the vagus nerve, which communicates with specific parts of the brain such as the hypothalamus. The vagus nerve receives information that digestive processes are happening in the gut and delivers advice directly to the brain as to what to do about food.

Hormonal secretions can control either the suppression of appetite or an increase in appetite in what are called anorexigenic and orexigenic reactions, respectively. Our brains get advice from these anorexigenic and orexigenic reactions. In one case, the brain is advised to stop eating (or at least slow down), and in the other the brain is advised to seek food. The anorexigenic reaction can occur as a result of many stimuli. One of these is the presence of lipids and fats in the duodenum. Lipids and fats are the products of food being digested in the stomach and intestines. In essence, the lipids and fats flip a switch in the cells of the intestinal lining to produce the hormone cholecystokinin. This hormone influences many aspects of appetite control, but its primary role is to interact with the neurons of the vagus nerve in the digestive tract. Once this happens, the gallbladder and pancreas release bile to degrade the fats and lipids, and the vagus-nerve neurons become primed for interactions with other hormones that then communicate advice to the brain. Because of its early involvement in this chain reaction, cholecystokinin is a very important hormone in the gut-brain axis.

Glucose in the intestines is another sign that food has been ingested, and high amounts of it indicate to the brain that regulation of appetite is needed. The presence of glucose, then, triggers the production of another hormone, insulin, which has a cascade of effects on the body and brain. There are several other hormones involved in the physiological process of advising the brain about satiety, and the regulatory system involving these hormones is complex.

Among the orexigenic hormones are ghrelin, which is secreted by the gut, and leptin, which is secreted by fat or adipose tissue. These two orexigenic hormones are transmitted to the hypothalamus to work synergistically to influence appetite. Ghrelin levels will increase before dining and decrease after food has been ingested. Whereas it is an important small molecule involved in the control of appetite, ghrelin has a broad distribution in other tissues and hence is thought to have far-reaching effects in the body.

With all of the complexity that is involved in the hormonal actions in the gut and brain, it is no wonder that changing the ecology of our guts can disrupt, or improve, many aspects of our physiology. The gut cells express some of the twenty thousand or so genes in the human genome. But the millions of bacteria in the stomach can express hundreds—even thousands—of different kinds of genes in the gut, far more than the average person does. Microbes in our guts affect our health in profound ways.

Our growing awareness of the importance of the microbiome has opened our eyes to gaps in our current approaches to disease management. Consider obesity, for instance. This disease as defined by the American Medical Association causes all kinds of systemic problems, which lead to poor health and to a shortened lifespan. But if we told you that obesity also might be involved in cancer, and that the ultimate reason cancer is a problem with obese people is bacterial, would that make sense? It turns out that studies in rodents have shown that obesity is controlled by genetic factors coupled with high-fat diets. In these rodents, obesity leads to altered communities of gut microbes. In other words, being obese means that certain kinds of microbes take over the gut. Specifically, the microbiota of an obese mouse's gut tends to harbor more species from the bacterial genus called *Clostridium*. This kind of bacterium likes bile acids that are produced copiously in the stomach to digest food. But these bacteria use a set of chemical reactions coded in their genomes to break down the bile—in short, they transform it into deoxycholic acid or DCA, which is a carcinogen. The higher concentration of *Clostridium* in the guts of obese mice makes for a toxic environment for the cells of the guts of these rodents and causes what is called senescence-associated secretory phenotype, which is associated with tumor growth.

In another compelling example of the complexity of neurohormonal interactions in the gut-brain axis, researchers in the past five years have used germfree mice to show that stress and depression are tightly tied to the microbiome. Depression is associated with an altered chemistry of what is known as the HPA (hypothalamic-pituitary-adrenal) axis. These three regions of the body are intricately involved in regulating a wide variety of behaviors and brain function, including mood. When communication among these major regulative regions of the body is altered, depressive states in the influenced individual may arise. In 2004, Nobuyuki Sudo and colleagues showed the first microbial link in the gut-brain axis. They used germfree bubble mice and mice with controlled microbiomes to examine the impact of the microbial community on stress. The stress level of the mice with different microbiomes was measured and the brain chemistries of the mice were then compared. Germfree mice were in general less stressed out than mice with diverse microbiomes. Specifically, ACTH or corticotropin and corticosterone, which are involved in the suppression of stress and anxiety in mice, were both expressed in elevated levels along the HPA axis. These results led to the hypothesis that the microbiome influences the hormones secreted in the HPA axis and hence the mouse's stress level.

Here is where the fun started in this experiment. To test their ideas, Sudo and colleagues performed some addition and subtraction experiments along the lines of Koch's hypothesis. They fed the mice with normal microbiomes two strains of the well-known gut species *E. coli*—one strain with pathogenic genes and one without—and asked if there was a change in stress response when the two strains were compared. The results? The mice with the benign strain of *E. coli* responded much like the mice in the germfree experiment, while mice with the pathogenic strain showed increased stress and decreased levels of ACTH and corticosterone. Next, the germfree mice were fed, at three different stages in their development, the feces of the mice that had had the pathogenic *E. coli*. When the germfree mice were fed the pathogenic fecal matter early in their development, their ACTH and corticosterone levels were suppressed, and they responded poorly to the stress tests. Feeding feces to adolescent mice, however, did not result in a change in ACTH or corticosterone levels, and the mice had a normal stress response, and adult mice fed lowered ACTH feces also expressed normal stress responses. This last experiment demonstrates that the microbes in the poop of stressed mice can transmit the stress response only in the early stages of infancy, when the gut microbiome is being established.

In another experiment, germfree bubble mice were fed a specific kind of microbe, *Lactobacillus rhamnosus,* and their behavior was compared to that of other mice that lacked this microbe. What's so special about *L. rhamnosus?* This interesting microbe usually doesn't reside inside us in very large quantities, even though it is a tough species that can withstand high pH and hence can live in the harsh environment of our stomachs, which are secreting bile and all kinds of acids. *L. rhamnosus* has been implicated as a beneficial regulator in all kinds of other bacterial and viral infections and has been used as a probiotic in many instances involving human health. For instance, it is used as a remedy for children with rotaviral infections that cause diarrhea and for children with atopic dermatitis disorders like eczema. So the good news is that this microbe is a good guy. The bad news is that we know very little about how it works its magic. Recently though, experiments in which bubble mice were fed this species began to illuminate its modus operandi.

One of the tests used by psychologists who work with mice to test their level of stress is the forced swim test. Naïve and healthy mice, when placed in water in a test chamber, will demonstrate climbing behavior (see Figure 6.7). Mice with perturbed behavior or health, by contrast, will simply give up and float (if a mouse ever dips below the water level it is removed immediately, so no harm actually comes to the mouse). Psychologists have used this test as an indicator of

Figure 6.7. The forced swim test (a mouse experiment indicating normal and abnormal responses to stress). A mouse will normally respond to the stress of being placed in a water-filled vessel by struggling to keep afloat (*left*). A mouse with an abnormal response to stress, however, will "give up" easily (*right*).

"depression" or of anxiety in mice after some perturbation. It turns out that mice become depressed just as humans do when they are placed under extreme physical or psychological stress, and stressed mice will give up after only a very short period when placed into the forced swim test chamber. The forced swim test is used widely to characterize the efficacy of antidepressant medications. So for instance, a normal mouse will be fed an appropriate dosage of an antidepressant, then placed into the chamber, and if it exhibits climbing behavior for a long period of time the drug is assumed to have worked at mitigating the stress of the forced swim. So what happens to bubble mice that are fed *L. rhamnosus* and then forced to swim in the chamber? They perform significantly better than mice without the microbe. Because the *L. rhamnosus* is the only difference between the experimental group and the controls, the presence of the microbe in the gut is assumed to be the mitigating factor. The *L. rhamnosus*–fed bubble mice also showed a higher level of expression in their brains of receptors for gamma-aminobutyric acid (GABA), a major neurotransmitter in the brain.

These results might be somewhat circumstantial in a court of law, but the real clincher is that if you give two groups of mice with normal microbiomes *L. rhamnosus* and then sever the vagus nerves of one group, the group with interrupted vagus nerves gives up much more quickly than the group with intact vagus nerves. This result means not only that the vagus nerve is critical for transmitting the information about the gut to the brain, but also, and more important, that this information about what is in the gut is important to the brain.

The major impact of *L. rhamnosus* on the gut microbiome is to regulate, like a bouncer, which other microbes are present. So if using a microbe to regulate the makeup of the microbiome works, then might antibiotics accomplish the same goal? Antimicrobial treatment, as we showed earlier, can allow researchers to make mice with very specific microbial gut communities, and such mice show lowered levels of anxiety too. (Treatment of bubble mice with antimicrobials has no effect, too, which reinforces the idea that the antibiotics are influencing the composition of the microbiome and that this behavioral effect is truly initiated or mediated by the microbiome.)

What is it about the presence of microbes in the gut that triggers the signals to the gut-brain axis? A change in microbial community structure might have several effects on the gut, but perhaps the most significant is related to inflammation. In the digestive tract, the permeability of the lining of the gut is critical for regulating the number and kinds of microbes living there. Stress increases the permeability of the gut membrane, and this in turn influences the gut microbiota. Recall that the gut mucosal membrane serves as a barrier to microbes, and there is a pretty effective system of immune cells that can take care of any microbes sneaking through. If the permeability of the gut lining is altered, though, this defense system is breached: large numbers of microbes can traverse the mucosal layer and start to interact with the immune cells and, equally important, interact with cells of the central nervous system. If the gut microbiota are altered with probiotics or with antibiotic treatment, then inflammation can often be stymied and the permeability of the gut membrane lowered, resulting in responses by the animal that reflect less neurological stress. Disorders like irritable bowel syndrome (IBS) or inflammatory bowel disease (IBD) are now being characterized in the context of inflammation and its impacts on the nervous system. It is no wonder that these disorders and Crohn's disease, another disorder of the bowel caused by inflammation, are also now routinely codiagnosed with psychological problems.

The influence of the inhabitants of the gut on the brain can be traced all the way back to infancy. By studying the gut microbiota of infant bubble mice, which are raised without normal microbiomes or without the presence of any microbes, researchers have shown an important role for the gut microbiome in how the neural system and the brain develop. In short, germfree bubble mice have higher levels of several important neurotransmitters in their brains like dopamine, norepinephrine, and serotonin. More important, they show a higher level of neural plasticity and an upregulated expression of several brain development genes that are involved in neural plasticity. All of these observations are linked to behavioral differences, because the germfree mice show lower levels of anxiety than do mice with manipulated, but normal, gut microbiomes. These results suggest a relationship between the microbiome and brain development.

Researchers have also begun to develop model systems to study autism spectrum disorder and the possible effects of the microbiome on this syndrome. Much work is still required to identify the possible role of the microbiome, but early research suggests a link between this often devastating disorder and the influence of the microbiome on brain development. For instance, researchers have developed mouse lines that show some of the neurological and behavioral symptoms of autism spectrum disorder. Using an approach called the maternal immune activation (MIA) model, the mouse mother's immune system is stimulated with a virus, and the subsequent effects on the mother and her offspring are observed. Such mice show at least three symptoms of autism: they are less vocal, less social, and show repetitive behaviors.

How the microbiome impacts this mouse model is currently under study. It looks as though the virus that stimulates maternal immune activation creates a dysbiosis of the gut, which lends an advantage to a genus of microbe called *Clostridium,* which produces a metabolite that in turn might pose a problem for healthy neural development. The researchers who use this mouse model suggest that probiotic treatment of the MIA mice with *Bacillus fragilis* can reverse the dysbiosis and stymie the autisticlike symptoms of MIA. Although this line of research is exciting and might eventually provide a possible therapy for autism spectrum disorders in humans, bear in mind, first, that the mouse model has often come up short in efforts to improve human health. In addition, any human therapy will involve the manipulation of a complex ecology involving many species of microbe and genes not only from ourselves, but also from hundreds of these microbial species. The strategy of manipulating whole-human microbiome ecologies is important and novel, but it is still in its infancy.

Epilogue

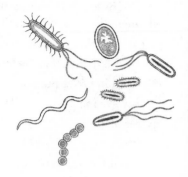

In the movie *The Princess Bride*, Buttercup, a beautiful young girl, is kidnapped and taken to marry a prince whom she doesn't love. On their way to the wedding, Buttercup and her captors, a trio of outlaws (a giant, an expert swordsman, and their leader), are pursued by a man in black. When the man in black catches up with them, he defeats the expert swordsman and the giant, then confronts the kidnappers' leader, Vizzini. The two men agree to a challenge of wits. The man in black pours iocane (a poison) into a goblet of wine, turns his back, and mixes up the goblets. He then places one of the goblets in front of Vizzini.

> MAN IN BLACK: All right. Where is the poison? The battle of wits has begun. It ends when you decide and we both drink, and find out who is right . . . and who is dead.
>
> VIZZINI: But it's so simple. All I have to do is divine from what I know of you: Are you the sort of man who would put the poison into his own goblet or his enemy's? Now, a clever man would put the poison into his own goblet, because he would know that only a great fool would reach for what he was given.

I am not a great fool, so I can clearly not choose the wine in front of you. But you must have known I was not a great fool, you would have counted on it, so I can clearly not choose the wine in front of me.

MAN IN BLACK: You've made your decision then?

The dialogue continues for another two minutes with Vizzini going back and forth, trying to reason the location of the poisoned cup. Then Vizzini picks a goblet, leaving the man in black the one he believes has the iocane in it. They both take a sip, and Vizzini snootily claims victory. Then he keels over, dead.

Our interactions with pathogens are a lot like Vizzini's with the iocane in the goblet. We are continually trying to stay ahead of microbes that do us harm by trying to outsmart them, even as they adapt to our best new strategies. This game is what scientists call an arms race, and we have been involved in arms races with microbial pathogens for ages.

We can make another point about microbial pathogens with this story from *The Princess Bride,* for it turns out that the man in black actually put iocane in both goblets of wine, and survived his drink because over the years he had developed a resistance to the poison. The moral of the story for us is that we can try to think of ways to avoid or kill pathogens, but they always will find a novel way to outwit us. Perhaps another strategy for managing pathogens might be like the one the man in black opted for: coexist with them by finding a way for them not to harm us. This is a huge task, and one we can approach only by understanding the nature of pathogenesis and how our bodies fight off invading microbes. We will have to reframe our views on sickness and health in a larger context that can include in our perspective what we know—and what we will come to know—about microbiomes, microbial-human interactions, and microbial ecology.

Recent discoveries about the complexity of microbiomes indicate that some of the usual frameworks for thinking about them are not robust enough. Although the one microbe, one disease paradigm has been very useful ever since microbes were seen as the causes of certain disorders, and Koch's postulates have been immeasurably helpful for understanding infectious diseases, these older paradigms of medical microbiology need to be rethought, or at least enhanced, given these complex ecological interactions. Much like the pioneering work that naturalists undertook in the early development of modern biology, modern medical microbiologists' path-breaking efforts to census and characterize the many organisms that live in and on us are vital because the species diversity and composition of a microbiome have become cornerstones in the study of infectious disease and human health.

In this next stage of understanding the human body, natural history is again at the forefront. Our modern lives are so different from our ancestors' lives in terms of diet, hygiene, antibiotics, immunization, and our ability to move about the globe that we are continually disrupting those microbiomes that have coevolved with us and other species. But scientists and medical professionals are not the only ones who need to be better educated in the ecological wonders of the microbial world: all humans do. For most of the microbes we encounter each day are beneficial to us, and to ignore or forget that fact—or to misjudge the effects of the antibiotics or antimicrobial compounds that have become seemingly ubiquitous in the modern world—could be very detrimental to our individual health and to the health of our species. Our very survival may well depend on understanding, and respecting, the ecology and evolutionary context of the microbiomes in and on us.

Glossary

Adaptive immune system: The subsystem of the vertebrate immune system that involves white blood cells that recognize and respond to antibodies bound to the surface of specific pathogens. Unlike the innate immune system, the adaptive—or "acquired"—immune system also involves the production of memory cells that can attack the pathogen much more quickly and on a larger scale should the organism become exposed again in the future.

Agar: A gelatin-like substance used to create nutrient-rich surfaces on which to grow microorganisms, particularly bacteria.

Alpha diversity: A measure of biological diversity within a given habitat, calculated as the number of species present within that habitat.

Amino acids: The basic building blocks of proteins. Amino acids are distinguished from each other by their side chains, a feature in their basic structure. There are twenty common side chains that make up the repertoire of amino acids in living organisms.

Antibody: A large, Y-shaped macromolecule used by the eukaryotic immune system to recognize and remove non-self particles. An antibody is also called an immunoglobulin (Ig). Antibodies work by binding to parts of antigens on the non-self particle. Through random combinations of gene segments, an organism's immune system can produce millions of subtle variations in their antibodies, which increases the likelihood that the body can detect a pathogen and mount a targeted immune response.

Antigen: A foreign structure or particle that can be recognized by an antibody based on its molecular shape or structure.

Antimicrobial: A compound that kills bacteria or prevents them from dividing into new cells.

Archaea/Archaebacteria: One of the three great domains of life. Discovered by Carl Woese in the 1980s using sequencing approaches, Archaea are single-celled organisms without a nuclear membrane around their genetic material, and they are distinguished from organisms in the other two domains by the structure of their cell membranes. Many but not all Archaea live in extreme environmental conditions. It is difficult to estimate the number of archaeal species alive on the planet, but the number certainly exceeds the eukaryotic number of two million.

Autoclave: A vessel sterilization device in which the temperature and pressure are raised in order to kill bacteria and viruses both in and on objects placed inside it.

Bacteria: One of the three great domains of life. Bacteria lack a nuclear membrane around their genetic material, and comprise the most abundant domain of life on the planet. It is difficult to estimate the number of species in the domain Bacteria, but some estimates are between ten million and a hundred million. The word "bacteria" (lower case) has also been used as a generic term to describe any microbe without a nucleus, though we suggest that this definition or notion of "bacteria" be dropped.

Bacterium: A single bacterial cell.

Beta diversity: A measure of biological diversity that reflects the uniqueness of a given habitat. Two places are said to have a high beta diversity if they have very few species in common.

Cell membrane: The phospholipid bilayer that separates the intracellular components from the environment.

Chromosome: A structure made up of deoxyribonucleic acid (DNA). In single-celled organisms without a nucleus, there is usually a single circular chromosome per genome. In Eukarya, chromosomes also have proteins called histones, as well as non-histone proteins, associated with them. These eukaryotic chromosomes are found in the nucleus and are usually linear. Chromosomes carry the genes of an organism.

Cloning vector: An artificially created plasmid used to introduce a segment of DNA into bacteria. Most cloning vectors contain genes for antibiotic resistance, which is a useful attribute for researchers trying to selectively screen the bacteria that have been transformed on an antibiotic-infused agar plate and thus detect those that possess plasmid DNA.

Codon: A three-nucleotide-long entity in DNA and RNA that codes for an amino acid in a protein or for the signal to stop protein synthesis. There are sixty-four possible ways that four bases (G, A, T, and C) can be arranged in triplets, so there are sixty-one possible codons in the standard DNA to code for the twenty amino acids and three codons which signal for translation to stop. Codons are the basic entities in the genetic code.

DNA: Deoxyribonucleic acid. A long chain molecule made up of deoxyribonucleotides and usually found in a double-stranded conformation. There are four major nitrogenous bases in DNA—guanine (G), cytosine (C), thymine (T), and adenine (A). Genes are comprised of DNA.

DNA barcode: A short DNA sequence that is used in biology as an identifier for a species. In animals, the cytochrome oxidase (COI) gene from the mitochondrial genome is used as a barcode. For Bacteria and Archaea, the 16S ribosomal RNA has been used as a DNA barcode.

Eukarya: One of the three great domains of life. Eukarya, also called eukaryotes, have a nuclear membrane around their genetic material and can be both unicellular and multicellular. Animals are Eukarya, and so are plants and fungi. Single-celled eukaryotes such as algae, amoebae, and pathogens such as *Plasmodium* also exist and are collectively called protists. There are roughly two million eukaryotic species alive today.

Evolution: The process of change in populations of organisms via descent with modification.

Fungus/fungi: A group of eukaryotic organisms with a single common ancestor, distinguished from other eukaryotes by having chitin in their cell walls. Fungi include yeasts, molds, and mushrooms, and they are more closely related to animals than to plants.

Gene: A stretch of DNA (or RNA) that codes for a biological structure. Most genes code for proteins, though some code for structures made of RNA (like ribosomal RNAs). Genes can be looked at as the basic units of heredity.

Habitat: The specific environment where a species of living organism is found.

Horizontal gene transfer: The process whereby an organism obtains genetic material in a manner other than its traditional mode of reproduction. Bacteria have three basic ways of engaging in horizontal gene transfer: (1) bacterial cells can take up free DNA from the environment, which is called transformation; (2) viruses can move DNA between host bacterial cells in a process called transduction; and (3) bacterial cells can transfer plasmids between themselves in a process known as conjugation.

Immunity/immune system: The cells and processes that an organism uses to protect itself from foreign bodies and organisms, including disease-causing pathogens, based on the ability of that organism's cells to recognize self and non-self cells and structures. The vertebrate immune system is usually broken down into two subsystems: the innate immune system and the adaptive (or "acquired") immune system.

Innate immune system: The subsystem of the vertebrate immune system that is not pathogen-specific and that utilizes for defense basic surface barriers, such as skin; inflammation at the site of infection; immune cells that destroy or consume any foreign or damaged cells; and a biochemical complimentary process that increases the effectiveness of an immune response.

Lipid: A biomolecule important in living organisms. The chemical definition of a lipid is that it is soluble in alcohol, but insoluble in water. The major chemical components of lipids are carbon, hydrogen, and oxygen.

Lymph: The fluid, composed partly of white blood cells, that circulates within the lymphatic system of an organism.

Lymphocytes: White blood cells that comprise a portion of the vertebrate immune system. They include the natural killer cells, T cells, and B cells.

Macrophages: The white blood cells that are capable of phagocytosis— the engulfing of other cells and debris in the body—which is an essential task of the immune system.

Microbiome: The type and number of microbes in a specific area such as a part of the human body. Microbiomes can vary temporally, and the term "microbiome" can also refer to the collection of genes present in the microbes associated with an organism.

Natural selection: A popular theory to explain many evolutionary patterns. Developed by Charles Darwin and Alfred Russell Wallace, it refers to the differential survival and reproduction of some members of a species.

Nucleotide: The basic building blocks of nucleic acids (RNA or DNA). There are three basic parts of a nucleotide—a phosphate tail, a sugar ring, and a nitrogenous-ringed base side chain.

Nucleus: A membrane-enclosed body found inside of eukaryotic cells that encases the genome.

Pathogen: Any agent that can cause disease.

PCR: Polymerase chain reaction. The PCR technique has become the most important tool in molecular biology. Invented by Kerry Mullis in 1984, it allows for the direct and targeted amplification of short DNA sequences from the genomes of organisms.

Petri dish/Petri plate: A shallow, covered dish made of glass or plastic into which agar is poured and microorganisms are grown.

Phylotype: A term used to distinguish lineages of microorganisms, which are more difficult to classify as separate species than eukaryotic, particularly multicellular, organisms.

Plasmid: An extra-chromosomal piece of DNA, usually circular in shape. Plasmids (and thus the genes they contain) can be exchanged between bacteria during horizontal gene transfer and may be artificially generated for the purposes of cloning.

Polymerase: A biomolecular enzyme that can generate a long-chain nucleic acid from a DNA or RNA template.

Prokaryote: A term used to describe single-celled organisms without nuclei (Archaea and Bacteria). We point out that this term has been relegated to the status of "non-biological" due to the lack of exclusive common ancestry of the two domains.

Protease: An enzyme that is capable of proteolysis, or the breaking down of proteins into amino acids or smaller polypeptides. Proteases are used in digestion and other bodily processes.

Protein: A long-chain biomolecule made up of amino acids. Proteins can be structural or enzymatic.

Ribosome: A cellular component with two subunits (large and small) that enables the translation of RNA into a protein. Ribosomes are made up of several structural RNAs, most notably the large-subunit RNA (also known as the 28S or 23S subunit) and the small-subunit RNA (also known as the 18S or 16S subunit), as well as several proteins called ribosomal proteins.

RNA: Ribonucleic acid. A long-chain molecule made up of ribonucleotides and usually found in a single-stranded conformation. RNA is composed of four types of nucleotides—guanine (G), cytosine (C), uracil (U), and adenine (A). Messenger RNA, the material that allows a cell to translate DNA to protein, is made up of RNA.

Sputum: The mucus that is present in the lungs and lower airways.

Sterilization: The process whereby microbes are completely removed or killed in a solution or on a surface.

Sugar: A molecule that includes carbon, hydrogen, and oxygen usually in a short chain. In general, sugars are sweet-tasting carbohydrates.

Swab: A sample of the microbes on a surface, often obtained by wiping an applicator of cotton or synthetic material across that surface. Alternatively, the word swab can be used to describe the cotton applicator itself.

Taxon: A group of populations of organisms that biologists have determined form a distinct biological unit in nature.

Taxonomy: The practice of naming living organisms in biology.

Vaccine: A protective dose of a killed or weakened microorganism, used to "train" the immune system to recognize and render inert a pathogenic version of the same kind of microorganism should the host become exposed to this same agent in the future.

Virome: The collection of all of the viruses (and/or their genes) present in association with another organism.

Virus: An entity that is entirely dependent on the biological machinery of the host cells that it infects for its own replication. Viral particles are generally made up of three major elements—a genome made of nucleic acids, a protein coat encasing the genome, and an outer layer of lipids. Virus genomes can be composed of double-stranded DNA or RNA, or single-stranded DNA or RNA.

References and Further Reading

Preface

The history of life on our planet is a fascinating subject. One of the most influential books on the subject of evolution is of course Charles Darwin's *On the Origin of Species*. Richard Fortey's book *Life* (2011) is a particularly good place to start for the history of life on the planet. The history of our existence as a species is covered in many books, including a detailed description in DeSalle and Tattersall (2008). And a great introduction to medical microbiology can be found in Blaser's 2014 book *Missing Microbes*.

Blaser, Martin J. 2014. *Missing Microbes: How the Overuse of Antibiotics Is Fueling Our Modern Plagues*. New York: Henry Holt.

Darwin, Charles. 1968 (1859). *On the Origin of Species by Means of Natural Selection*. London: Murray.

DeSalle, Rob, and Ian Tattersall. 2008. *Human Origins: What Bones and Genomes Tell Us about Ourselves*. College Station: Texas A&M University Press.

Fortey, Richard. 2011. *Life: A Natural History of the First Four Billion Years of Life on Earth*. New York: Random House.

CHAPTER 1. What Is Life?

There are many primers on molecular biology and genomics in the literature (for example, DeSalle and Yudell 2004; Quackenbush 2011; Davies 2010) and these can be consulted for background in the science of genomics. There are also many excellent educational websites—such as The DNA Learning Center at http://www.dnalc.org and DNA Interactive at http://www.dnai.org—that can provide some background to the molecular biology that we discuss in this chapter. The early history of life and species on the Earth is discussed in Richard Fortey's *Life* (2011), Nick Lane's *Life Ascending* (2010), E. O. Wilson's *The Diversity of Life* (1999), and Norm Pace (2009), which provides a bacterial perspective. The potential impact of horizontal gene transfer on the bacterial tree of life has been discussed by Ford Doolittle (2009), and recent work on the very first branches of the tree of life is reviewed in articles by Williams et al. (2013) and Williams (2014). The references in this chapter to rooting the tree of life are in Gogarten et al. (1989) and Iwabe et al. (1989). Susumu Ohno's book *Evolution by Gene Duplication* and recent review articles on whole genome duplication are a good place to delve into the phenomenon of duplication (Dehal and Boore 2005; Meyer et al. 2005; DeBodt et al. 2005). Eugene Koonin (2003) and Arcady Mushegian have discussed the role of LUCA in our understanding of evolution of life on the planet. The late Lynn Margulis, too, wrote prolifically on symbiosis of eukaryotes and on the evolution of the eukaryotic cell, and we cite one of her books that explains her take on this phenomenon (Margulis 2008).

Davies, Kevin. 2010. *The $1,000 Genome: The Revolution in DNA Sequencing and the New Era of Personalized Medicine.* New York: Simon and Schuster.

De Bodt, Stefanie, Steven Maere, and Yves Van de Peer. 2005. "Genome Duplication and the Origin of Angiosperms." *Trends in Ecology & Evolution* 20, no. 11:591–597.

Dehal, Paramvir, and Jeffrey L. Boore. 2005. "Two Rounds of Whole Genome Duplication in the Ancestral Vertebrate." *PLoS Biology* 3, no. 10:e314.

DeSalle, Rob, and Michael Yudell. 2004. *Welcome to the Genome: A User's Guide to the Genetic Past, Present, and Future.* New York: Wiley.

Doolittle, W. Ford. 2009. "The Practice of Classification and the Theory of Evolution, and What the Demise of Charles Darwin's Tree of Life Hypothesis Means for Both of Them." *Philosophical Transactions of the Royal Society B: Biological Sciences* 364, no. 1527:2221–2228.

Fortey, Richard. 2011. *Life: A Natural History of the First Four Billion Years of Life on Earth.* New York: Random House.

Gogarten, J. P., et al. 1989. "Evolution of the Vacuolar H+−ATPase—Implications for the Origin of Eukaryotes." *Proceedings of the National Academy of Sciences of the United States of America* 86:6661–6665.

Iwabe, N., et al. 1989. "Evolutionary Relationship of Archaebacteria, Eubacteria, and Eukaryotes Inferred from Phylogenetic Trees of Duplicated Genes." *Proceedings of the National Academy of Sciences of the United States of America* 86:9355–9359.

Koonin, Eugene V. 2003. "Comparative Genomics, Minimal Gene-Sets and the Last Universal Common Ancestor." *Nature Reviews Microbiology* 1, no. 2:127–136.

Lane, Nick. 2010. *Life Ascending: The Ten Great Inventions of Evolution.* London: Profile Books.

Lombard, Jonathan, Purificación López-García, and David Moreira. 2012. "The Early Evolution of Lipid Membranes and the Three Domains of Life." *Nature Reviews Microbiology* 10, no. 7:507–515.

Margulis, Lynn. 2008. *Symbiotic Planet: A New Look at Evolution.* New York: Basic Books.

Meyer, Axel, and Yves Van de Peer. 2005. "From 2R to 3R: Evidence for a Fish-Specific Genome Duplication (FSGD)." *Bioessays* 27, no. 9:937–945.

Mushegian, Arcady. 2007. "Gene Content of LUCA, the Last Universal Common Ancestor." *Frontiers in Bioscience: A Journal and Virtual Library* 13:4657–4666.

Ohno, Susumu. 1970. *Evolution by Gene Duplication.* London: George
    Allen & Unwin.

Pace, Norman R. 2009. "Mapping the Tree of Life: Progress and Pros-
    pects." *Microbiology and Molecular Biology Reviews* 73:565–576.

Quackenbush, John. 2011. *Curiosity Guides: The Human Genome.*
    Charlesbridge.

Williams, Tom A. 2014. "Evolution: Rooting the Eukaryotic Tree of
    Life." *Current Biology* 24, no. 4:R151–R152.

Williams, Tom A., et al. 2013. "An Archaeal Origin of Eukaryotes
    Supports Only Two Primary Domains of Life." *Nature* 504,
    no. 7479:231–236.

Wilson, Edward O. 1999. *The Diversity of Life.* New York: Norton.

CHAPTER 2. What Is a Microbiome?

The brief history of microbiology that we provide in Chapter 2 can
be expanded by consulting a classic treatment of the subject by Milton Wainwright and the Nobel Prize–winning microbiologist Joshua
Lederberg (Wainwright and Lederberg 1992). The modern formulation of Koch's postulates from a molecular perspective is discussed in
Falkow (1988), and for an accessible discussion of the history of the
discovery of disease microbes we point the reader to Paul de Kruif's
Microbe Hunters, first published in 1926. The classical approaches to
characterizing microbes in nature using DNA sequencing are discussed
in detail in Woese et al. (1990), Pace (2009), and Wu et al. (2013). The
basics of microbial ecology and how these relate to the microbiome are
discussed in more detail in Fierer et al. (2012), and a historical treatment of infectious disease and microbes can be found in Casanova and
Abel (2013). For a primer of DNA sequencing of microbiomes we direct
the reader to Chapter 1 and to Tringe and Rubin (2005). DNA barcoding has a large literature and can be accessed in Hebert, Ratnasingham,
and Waard (2003) and in Goldstein and DeSalle (2011). The early meta-
genome studies we describe in the text are from Schmidt, DeLong, and
Pace (1991), Venter et al. (2004), Ward et al. (1990), and Giovannoni
et al. (1990). The medical approach using the microbiome mentioned

in the text can be attributed to Relman and Falkow (2001) and Relman (2002). The growth of our understanding of bacterial diversity, illustrated in the tree figure in this chapter, was adapted from a figure drawn by Norm Pace and Kirk Harris (http://forms.asm.org/microbe/index .asp?bid=32571). For the 2011 pond water study led by Thomas Richards, see Jones et al. (2011). For more on some of the novel strategies embodied in next-generation DNA sequencing (NGS) approaches, see Metzger (2010) and Mardis (2008). Details on the Human Microbiome Project (HMP) can be found in several papers published in high-impact journals by the Human Microbiome Project Consortium and by Blaser et al. (2013). A review of next-generation approaches to characterizing microbiomes can be found in Wessel et al. (2013).

Blaser, Martin, et al. 2013. "The Microbiome Explored: Recent Insights and Future Challenges." *Nature Reviews Microbiology* 11, no. 3:213–217.

Casanova, Jean-Laurent, and Laurent Abel. 2013. "The Genetic Theory of Infectious Diseases: A Brief History and Selected Illustrations." *Annual Review of Genomics and Human Genetics* 14:215–243.

de Kruif, P. 2002 (1926). *Microbe Hunters.* Houghton Mifflin Harcourt.

Falkow, S. 1988. "Molecular Koch's Postulates Applied to Microbial Pathogenicity." *Reviews of Infectious Diseases* 10, supp. 2:S274–S276.

Fierer, Noah, et al. 2012. "From Animalcules to an Ecosystem: Application of Ecological Concepts to the Human Microbiome." *Annual Review of Ecology, Evolution, and Systematics* 43:137–155.

Giovannoni, Stephen J., et al. 1990. "Genetic Diversity in Sargasso Sea Bacterioplankton." *Nature* 345, no. 6270:60–63.

Goldstein, Paul Z., and Rob DeSalle. 2011. "Integrating DNA Barcode Data and Taxonomic Practice: Determination, Discovery, and Description." *Bioessays* 33, no. 2:135–147.

Hebert, Paul D. N., Sujeevan Ratnasingham, and Jeremy R. de Waard. 2003. "Barcoding Animal Life: Cytochrome C Oxidase Subunit 1 Divergences among Closely Related Species." *Proceedings of the*

Royal Society of London. Series B: Biological Sciences 270, supp. 1:S96–S99.

Human Microbiome Project Consortium. 2012. "A Framework for Human Microbiome Research." *Nature* 486, no. 7402:215–221.

———. 2012. "Structure, Function and Diversity of the Healthy Human Microbiome." *Nature* 486, no. 7402:207–214.

Jones, Meredith D. M., et al. 2011. "Discovery of Novel Intermediate Forms Redefines the Fungal Tree of Life." *Nature* 474 (June 9): 200–203.

Jumpstart Consortium Human Microbiome Project Data Generation Working Group. 2012. "Evaluation of 16S rDNA-based Community Profiling for Human Microbiome Research." *PLoS One* 7, no. 6:e39315.

Mardis, Elaine R. 2008. "Next-Generation DNA Sequencing Methods." *Annual Review of Genomics and Human Genetics* 9:387–402.

Metzker, Michael L. 2010. "Sequencing Technologies—The Next Generation." *Nature Reviews Genetics* 11, no. 1:31–46.

Pace, Norman R. 2009. "Mapping the Tree of Life: Progress and Prospects." *Microbiology and Molecular Biology Reviews* 73:565–576.

Relman, D. A. 2002. "New Technologies, Human-Microbe Interactions, and the Search for Previously Unrecognized Pathogens." *Journal of Infectious Disease* 186:S254–S258.

Relman D. A., and S. Falkow. 2001. "The Meaning and Impact of the Human Genome Sequence for Microbiology." *Trends in Microbiology* 9:206–208.

Schmidt, T. M., E. F. DeLong, and N. R. Pace. 1991. "Analysis of a Marine Picoplankton Community by 16S rRNA Gene Cloning and Sequencing." *Journal of Bacteriology* 173, no. 14:4371.

Tringe, Susannah Green, and Edward M. Rubin. 2005. "Metagenomics: DNA Sequencing of Environmental Samples." *Nature Reviews Genetics* 6, no. 11:805–814.

Venter, J. C., et al. 2004. "Environmental Genome Shotgun Sequencing of the Sargasso Sea." *Science* 304:66–74.

Wainwright, Milton, and J. Lederberg. 1992. "History of Microbiology." *Encyclopedia of Microbiology* 2:419–437.

Ward, David M., Roland Weller, and Mary M. Bateson. 1990. "16S rRNA Sequences Reveal Numerous Uncultured Microorganisms in a Natural Community." *Nature* 345, no. 6270:63–65.

Wessel, Aimee K., et al. 2013. "Going Local: Technologies for Exploring Bacterial Microenvironments." *Nature Reviews Microbiology* 11, no. 5:337–348.

Woese, C. R., O. Kandler, and M. L. Wheelis. 1990. "Towards a Natural System of Organisms: Proposal for the Domains Archaea, Bacteria, and Eucarya." *Proceedings of the National Academy of Sciences of the United States of America* 87:4576–4579.

Wu, Dongying, et al. 2009. "A Phylogeny-Driven Genomic Encyclopaedia of Bacteria and Archaea." *Nature* 462, no. 7276:1056–1060.

CHAPTER 3. What Is On and Around Us?

A description of the human microbiome project (HMP) can be found in Chapter 2 and its references; see also Turnbaugh et al. (2007), Peterson et al. (2009), and Gevers et al. (2012). Handedness and the skin microbiomes are discussed by Fierer et al. (2008). For our discussion on what is on us, we used a lot of information from Grice et al. (2011a and 2011b), and for more on roller derby girls and bellybutton microbiomes, see Meadow et al. (2013) and Hulcr et al. (2012). The microbiomes of pregnancy and newborns are referred to in Aagaard et al. (2012). Subway microbiome references include Robertson et al. (2013), Dybwad (2012), Ki Youn (2011), and the PathoMap project in New York City (http://www.pathomap.org). The Louvre Museum study (Gaüzère et al. 2014), the University of Colorado study (Flores et al. 2011), houseomes, cellphone-omes, shoe-omes, and office-omes have been discussed in several publications (Hewitt et al. 2012; Dunn et al. 2013; Fierer et al. 2010) and on the Home Microbiome Project website (http://homemicrobiome.com). More on dogomes can be found in Rodrigues et al. (2013) and Fujimura et al. (2010).

Aagaard, Kjersti, et al. 2012. "A Metagenomic Approach to Characterization of the Vaginal Microbiome Signature in Pregnancy." *PloS One* 7, no. 6:e36466.

Dunn, Robert R., et al. 2013. "Home Life: Factors Structuring the Bacterial Diversity Found within and between Homes." *PloS One* 8, no. 5:e64133.

Dybwad, Marius, et al. 2012. "Characterization of Airborne Bacteria at an Underground Subway Station." *Applied and Environmental Microbiology* 78, no. 6:1917–1929.

Fierer, Noah, et al. 2008. "The Influence of Sex, Handedness, and Washing on the Diversity of Hand Surface Bacteria." *Proceedings of the National Academy of Sciences of the United States of America* 105, no. 46:17994–17999.

———. 2010. "Forensic Identification Using Skin Bacterial Communities." *Proceedings of the National Academy of Sciences of the United States of America* 107, no. 14:6477–6481.

Flores, Gilberto E., et al. 2011. "Microbial Biogeography of Public Restroom Surfaces." *PLoS One* 6, no. 11:e28132.

Fujimura, Kei E., et al. 2010. "Man's Best Friend? The Effect of Pet Ownership on House Dust Microbial Communities." *Journal of Allergy and Clinical Immunology* 126, no. 2:410.

Gaüzère, Carole, et al. 2014. "Stability of Airborne Microbes in the Louvre Museum over Time." *Indoor Air* 24, no. 1:29–40.

Gevers, Dirk, et al. 2012. "The Human Microbiome Project: A Community Resource for the Healthy Human Microbiome." *PLoS Biology* 10, no. 8:e1001377.

Grice, Elizabeth A., and Julia A. Segre. 2011a. "The Human Microbiome: Our Second Genome." *Annual Review of Genomics and Human Genetics* 13:151–170.

———. 2011b. "The Skin Microbiome." *Nature Reviews Microbiology* 9, no. 4:244–253.

Hewitt, Krissi M., et al. 2012. "Office Space Bacterial Abundance and Diversity in Three Metropolitan Areas." *PLoS One* 7, no. 5:e37849.

Hulcr, Jiri, et al. 2012. "A Jungle in There: Bacteria in Belly Buttons Are Highly Diverse, But Predictable." *PloS One* 7, no. 11:e47712.

Ki Youn, K. I. M., et al. 2011. "Exposure Level and Distribution Characteristics of Airborne Bacteria and Fungi in Seoul Metropolitan Subway Stations." *Industrial Health* 49:242–248.

Meadow, James F., et al. 2013. "Significant Changes in the Skin Microbiome Mediated by the Sport of Roller Derby." *PeerJ* 1:e53.

Peterson, Jane, et al. 2009. "The NIH Human Microbiome Project." *Genome Research* 19, no. 12:2317–2323.

Robertson, Charles E., et al. 2013. "Culture-Independent Analysis of Aerosol Microbiology in a Metropolitan Subway System." *Applied and Environmental Microbiology* 79, no. 11:3485–3493.

Rodrigues, Hoffmann A., et al. 2014. "The Skin Microbiome in Healthy and Allergic Dogs." *PloS One* 9, no. 1:e83197.

Sinkkonen, Aki. "Umbilicus as a Fitness Signal in Humans." *FASEB Journal* 23, no. 1:10–12.

Turnbaugh, Peter J., et al. 2007. "The Human Microbiome Project: Exploring the Microbial Part of Ourselves in a Changing World." *Nature* 449, no. 7164:804.

CHAPTER 4. What Is Inside Us?

The following papers will help readers to understand the approaches to characterizing the gut oral microbiome and the great diversity of microbes that are present in this cavity and the digestive tract. For Anne Tanner's study of dental caries, see Hughes et al. (2012). Quorum sensing and biofilms are discussed in Miller and Bassler (2001), Costerton et al. (1999), and Burmølle et al. (2014). Gut microbiomes are discussed in detail in Yatsunenko et al. (2012) and the Enterotest is introduced in Fillon et al. (2012). Other gut microbiome studies discussed in this chapter can be found in Chewapreecha (2014), Sommer and Bäckhed (2013), Everard (2013), Segata et al. (2012), and Kau et al. (2011). The fecal virome has been examined in Reyes et al. (2010). The vaginal microbiome has been reported on in Ma et al. (2012), Solt and Cohavy (2012), and Aagaard et al. (2012). The penis microbiome is described in

Price et al. (2010), Nelson et al. (2010), and Mändar (2013). Lung micro-biomes and "healthy smokers" are discussed further in Beck, Young, and Huffnagle (2012), Pragman et al. (2012), Dickson, Erb-Downward, and Huffnagle (2014), and Erb-Downward et al. (2011).

Aagaard, Kjersti, et al. 2012. "A Metagenomic Approach to Character-ization of the Vaginal Microbiome Signature in Pregnancy." *PloS One* 7, no. 6:e36466.

Aas, Jørn A., et al. 2005. "Defining the Normal Bacterial Flora of the Oral Cavity." *Journal of Clinical Microbiology* 43, no. 11:5721–5732.

———. 2008. "Bacteria of Dental Caries in Primary and Perma-nent Teeth in Children and Young Adults." *Journal of Clinical Microbiology* 46, no. 4:1407–1417.

Beck, James M., Vincent B. Young, and Gary B. Huffnagle. 2012. "The Microbiome of the Lung." *Translational Research* 160, no. 4:258–266.

Bik, Elisabeth M., et al. 2010. "Bacterial Diversity in the Oral Cavity of Ten Healthy Individuals." *ISME Journal* 4, no. 8:962–974.

Burmølle, Mette, et al. 2014. "Interactions in Multispecies Biofilms: Do They Actually Matter?" *Trends in Microbiology* 22, no. 2:84–91.

Chen, Tsute, et al. 2010. "The Human Oral Microbiome Database: A Web Accessible Resource for Investigating Oral Microbe Taxo-nomic and Genomic Information." *Database: The Journal of Biological Databases and Curation.* July 6.

Chewapreecha, Claire. 2014. "Your Gut Microbiota Are What You Eat." *Nature Reviews Microbiology* 12, no. 1:8.

Costerton, J. W., Philip S. Stewart, and E. P. Greenberg. 1999. "Bacte-rial Biofilms: A Common Cause of Persistent Infections." *Science* 284, no. 5418:1318–1322.

Darveau, Richard P. 2010. "Periodontitis: A Polymicrobial Disrup-tion of Host Homeostasis." *Nature Reviews Microbiology* 8, no. 7:481–490.

Dewhirst, Floyd E., et al. 2010. "The Human Oral Microbiome." *Jour-nal of Bacteriology* 192, no. 19:5002–5017.

Dickson, Robert P., John R. Erb-Downward, and Gary B. Huffnagle. 2014. "Towards an Ecology of the Lung: New Conceptual Models of Pulmonary Microbiology and Pneumonia Pathogenesis." *Lancet Respiratory Medicine* 2, no. 3:238–246.

Erb-Downward, John R., et al. 2011. "Analysis of the Lung Microbiome in the 'Healthy' Smoker and in COPD." *PLoS One* 6, no. 2:e16384.

Everard, A. 2013. "Cross-Talk between *Akkermansia muciniphila* and Intestinal Epithelium Controls Diet-Induced Obesity." *Proceedings of the National Academy of Sciences of the United States of America*. 13 May. doi:10.1073/pnas.1219451110.

Fillon, Sophie A., et al. 2012. "Novel Device to Sample the Esophageal Microbiome—The Esophageal String Test." *PloS One* 7, no. 9:e42938.

Ge, Xiuchun, et al. 2013. "Oral Microbiome of Deep and Shallow Dental Pockets in Chronic Periodontitis." *PloS One* 8, no. 6:e65520.

Hughes, Christopher, et al. 2012. "Aciduric Microbiota and Mutans Streptococci in Severe and Recurrent Severe Early Childhood Caries." *Pediatric Dentistry* 34, no. 2:16–23.

Kau, Andrew L., et al. 2011. "Human Nutrition, the Gut Microbiome and the Immune System." *Nature* 474, no. 7351:327–336.

Li, Jing, et al. 2013. "The Saliva Microbiome of Pan and Homo." *BMC Microbiology* 13, no. 1:204.

Liu, Cindy M., et al. 2011. "The Otologic Microbiome: A Study of the Bacterial Microbiota in a Pediatric Patient with Chronic Serous Otitis Media Using 16SrRNA Gene-Based Pyrosequencing." *Archives of Otolaryngology—Head & Neck Surgery* 137, no. 7:664–668.

Lowe, Beth A., et al. 2012. "Defining the 'Core Microbiome' of the Microbial Communities in the Tonsils of Healthy Pigs." *BMC Microbiology* 12, no. 1:20.

Ma, Bing, Larry J. Forney, and Jacques Ravel. 2012. "The Vaginal Microbiome: Rethinking Health and Diseases." *Annual Review of Microbiology* 66:371.

Mändar, Reet. 2013. "Microbiota of the Male Genital Tract: Impact on the Health of Man and His Partner." *Pharmacological Research* 69, no. 1:32–41.

Miller, Melissa B., and Bonnie L. Bassler. 2001. "Quorum Sensing in Bacteria." *Annual Reviews in Microbiology* 55, no. 1:165–199.

Nasidze, Ivan, et al. 2009. "Global Diversity in the Human Salivary Microbiome." *Genome Research* 19, no. 4:636–643.

Nelson, David E., et al. 2010. "Characteristic Male Urine Microbiomes Associate with Asymptomatic Sexually Transmitted Infection." *PLoS One* 5, no. 11:e14116.

Pragman, Alexa A., et al. 2012. "The Lung Microbiome in Moderate and Severe Chronic Obstructive Pulmonary Disease." *PLos One* 7, no. 10:e47305.

Price, Lance B., et al. 2010. "The Effects of Circumcision on the Penis Microbiome." *PLoS One* 5, no. 1:e8422.

Reyes, Alejandro, et al. 2010. "Viruses in the Faecal Microbiota of Monozygotic Twins and Their Mothers." *Nature* 466, no. 7304:334–338.

Segata, Nicola, et al. 2012. "Composition of the Adult Digestive Tract Bacterial Microbiome Based on Seven Mouth Surfaces, Tonsils, Throat and Stool Samples." *Genome Biology* 13, no. 6:R42.

Solt, Ido, and Offer Cohavy. 2012. "The Great Obstetrical Syndromes and the Human Microbiome—A New Frontier." *Rambam Maimonides Medical Journal* 3, no. 2.

Sommer, F., and F. Bäckhed. 2013. "The Gut Microbiota—Masters of Host Development and Physiology." *Nature Reviews Microbiology* 11:227–238.

Walter, Jens, and Ruth Ley. 2011. "The Human Gut Microbiome: Ecology and Recent Evolutionary Changes." *Annual Review of Microbiology* 65:411–429.

Yatsunenko, Tanya, et al. 2012. "Human Gut Microbiome Viewed across Age and Geography." *Nature* 486, no. 7402:222–227.

Zaura, Egija, et al. 2009. "Defining the Healthy." *BMC Microbiology* 9, no. 1:259.

CHAPTER 5. What Are Our Defenses?

Semmelweis and his great contribution to medicine can be found in his classic paper reproduced in 1981, as well as in the account by Adriaanse, Pel, and Bleker (2000). Our discussion of the evolution of the immune system referred to several papers on the immune systems of lower animals and the more recent evolution of the vertebrate immune system; see Rodríguez, López-Vázquez, and López-Larrea (2012); Hill and Artis (2009); Cooper and Herrin (2010); Lemaitre and Hoffman (2007); Bosch (2013); and Spoel and Dong (2012). There are many great websites that take on the description of the immune system; see, for example, http://www.niaid.nih.gov/topics/immunesystem/Pages/default.aspx, http://www.livescience.com/26579-immune-system.html, and http://medicalcenter.osu.edu/patientcare/healthcare_services/infectious_diseases/immunesystem/Pages/index.aspx. For more on Maurice Hilleman and vaccines, see Offit (2008). Horizontal gene transfer and tight adherence are discussed in detail in Syvanen (2012) and Planet et al. (2003). Camel dung is referenced from Damman et al. (2012). And for a video on the web showing phagocytic cells of the human immune systems persistently chasing down invading bacteria, see https://www.youtube.com/watch?v=KxTYyNEbVU4.

Adriaanse, Albert H., Maria Pel, and Otto P. Bleker. 2000. "Semmelweis: The Combat against Puerperal Fever." *European Journal of Obstetrics & Gynecology and Reproductive Biology* 90, no. 2:153–158.

Bosch, Thomas C. G. 2013. "Cnidarian-Microbe Interactions and the Origin of Innate Immunity in Metazoans." *Annual Review of Microbiology* 67:499–518.

Cooper, Max D., and Brantley R. Herrin. 2010. "How Did Our Complex Immune System Evolve?" *Nature Reviews Immunology* 10, no. 1:2–3.

Damman, Christopher J., et al. 2012. "The Microbiome and Inflammatory Bowel Disease: Is There a Therapeutic Role for Fecal Microbiota Transplantation?" *American Journal of Gastroenterology* 107, no. 10:1452–1459.

Hill, David A., and David Artis. 2009. "Intestinal Bacteria and the Regulation of Immune Cell Homeostasis." *Annual Review of Immunology* 28:623–667.

Lemaitre, Bruno, and Jules Hoffmann. 2007. "The Host Defense of Drosophila Melanogaster." *Annual Review of Immunology* 25:697–743.

Offit, Paul. 2008. *Vaccinated: One Man's Quest to Defeat the World's Deadliest Diseases.* New York: Harper Perennial.

Planet, Paul J., et al. 2003. "The Widespread Colonization Island of *Actinobacillus actinomycetemcomitans.*" *Nature Genetics* 34, no. 2:193–198.

Rodríguez, Ramón M., Antonio López-Vázquez, and Carlos López-Larrea. 2012. "Immune Systems Evolution." Pp. 237–251 in Carlos López-Larrea, ed., *Sensing in Nature.* New York: Springer.

Semmelweis, Ignaz Philipp. 1981. "Childbed Fever." *Review of Infectious Diseases* 3, no. 4:808–811.

Spoel, Steven H., and Xinnian Dong. 2012. "How Do Plants Achieve Immunity? Defense without Specialized Immune Cells." *Nature Reviews Immunology* 12, no. 2:89–100.

Syvanen, Michael. 2012. "Evolutionary Implications of Horizontal Gene Transfer." *Annual Review of Genetics* 46:341–358.

CHAPTER 6. What Is "Healthy"?

Human health and the microbiome is a broad topic and we have chosen to focus on a couple of the high-visibility problems that have been examined using the microbiome. For solid background information on how human genes interact with microbes from the evolutionary theory and human health perspectives, see Genovese et al. (2010), Honda and Littman (2012), Cho and Blaser (2012), Falush et al. (2003), Tamboli et al. (2004), and the work of Blaser (2006), who uses *Helicobacter pylori* to examine the complexity of microbial interactions in our bodies. Hawrelak and Myers (2004); Stecher, Maier, and Hardt (2013); Kamada et al. (2013); and Turnbaugh et al. (2006) discuss the gut microbiome and bacterial blooms as a phenomenon in the gut. Ma, Forney, and

Ravel (2012) introduce some important concepts for the understanding of bacterial-vaginosis-associated bacteria and explain how we need to rethink our formulation of pathogenesis based on the microbiology of the vagina. Zhao (2013), Sommer and Bäckhed (2013), and Sudo et al. (2004) discuss the role of the gut in influencing our brains. Finally, Foster and Neufeld (2013) and Romijn et al. (2008) define the gut-brain axis.

Blaser, Martin J. 2006. "Who Are We? Indigenous Microbes and the Ecology of Human Diseases." *EMBO Reports* 7, no. 10:956–960.

Cho, Ilseung, and Martin J. Blaser. 2012. "The Human Microbiome: At the Interface of Health and Disease." *Nature Reviews Genetics* 13, no. 4:260–270.

Falush, Daniel, et al. 2003. "Traces of Human Migrations in *Helicobacter pylori* Populations." *Science* 299, no. 5612:1582–1585.

Foster, Jane A., and Karen-Anne McVey Neufeld. 2013. "Gut-Brain Axis: How the Microbiome Influences Anxiety and Depression." *Trends in Neurosciences* 36, no. 5:305–312.

Genovese, Giulio, et al. 2010. "Association of Trypanolytic ApoL1 Variants with Kidney Disease in African Americans." *Science* 329, no. 5993:841–845.

Hawrelak, Jason A., and Stephen P. Myers. 2004. "The Causes of Intestinal Dysbiosis: A Review." *Alternative Medicine Review* 9:180–192.

Honda, Kenya, and Dan R. Littman. 2012. "The Microbiome in Infectious Disease and Inflammation." *Annual Review of Immunology* 30:759–795.

Kamada, Nobuhiko, et al. 2013. "Role of the Gut Microbiota in Immunity and Inflammatory Disease." *Nature Reviews Immunology* 13, no. 5:321–335.

Ma, Bing, Larry J. Forney, and Jacques Ravel. 2012. "The Vaginal Microbiome: Rethinking Health and Diseases." *Annual Review of Microbiology* 66:371.

Romijn, Johannes A., et al. 2008. "Gut-Brain Axis." *Current Opinion in Clinical Nutrition & Metabolic Care* 11, no. 4:518–521.

Sommer, Felix, and Fredrik Bäckhed. 2013. "The Gut Microbiota—Masters of Host Development and Physiology." *Nature Reviews Microbiology* 11, no. 4:227–238.

Stecher, Bärbel, Lisa Maier, and Wolf-Dietrich Hardt. 2013. "'Blooming' in the Gut: How Dysbiosis Might Contribute to Pathogen Evolution." *Nature Reviews Microbiology* 11, no. 4:277–284.

Sudo, Nobuyuki, et al. 2004. "Postnatal Microbial Colonization Programs the Hypothalamic—Pituitary—Adrenal System for Stress Response in Mice." *Journal of Physiology* 558, no. 1:263–275.

Tamboli, C. P., et al. 2004. "Dysbiosis in Inflammatory Bowel Disease." *Gut* 53, no. 1:1–4.

Travis, John. 2009. "On the Origin of the Immune System." *Science* 324, no. 5927: 580–582.

Turnbaugh, Peter J., et al. 2006. "An Obesity-Associated Gut Microbiome with Increased Capacity for Energy Harvest." *Nature* 444, no. 7122:1027–1131.

Zhao, Liping. 2013. "The Gut Microbiota and Obesity: From Correlation to Causality." *Nature Reviews Microbiology* 11, no. 9:639–647.

Index

A (adenine), 6–7
acidity. *See* pH
Acinetobacter spp., 91
ACTH (corticotropin, corticosterone), 199–200
Actinobacteria, 73, 74, 75, 85, 93, 95
adaptation. *See* variation and adaptation
adenine (A), 6–7
adenoids, 109–110
Aerococcus spp., 127–128
AIDS. *See* HIV/AIDS
Akkermansia muciniphila, 120, 188–190
Alcaligenes spp., 79
Alcanivorax spp., 79
alpha diversity, 50, 176–178
alpha-globin protein, 168
alveoli (alveolus), 140–141
amino acids, 6–11, 24
amoeba, 104
AMPs (antimicrobial peptides), 77, 184
Amsel and Nugent scores, 192–193
Anaerococcus spp., 127–128
anemia, 168–170
animalcules, 40
anorexigenic reactions, 197
antibiotics. *See* antimicrobials

antibodies (immunoglobulins, Ig), 119, 152–154, 184
antigen receptors, 152, 154
antigens, 128, 152–153, 154
antimicrobial peptides (AMPs), 77, 184
antimicrobials: biofilms and, 114; and bubble mice, 180–181, 202–203; and composition of the microbiome, 202–203; discovery of penicillin, 154–156; and health, 208; hydrozoans and, 145; innate immune system, 145–147; resistance to, 76–77, 157–163, 177; and viruses, 66; Waksman's research on, 156–157
apoptosis, 145, 150
appetite regulation, 175, 196–198
Archaea: in belly button microbiome, 85; cell characteristics from LUCA (last universal common ancestor), 25–30; detection and identification of (*see* detection and identification of microbes); genomes of, 19–25; numbers of, 30; as one of three major cell types, x, 2, 13, 17–19; in oral cavity microbiome, 104; and tree of life, x, 19(fig.), 20–21, 23–25; unique traits of, 33–34

"Archaea first" hypothesis, 24
Aristotle, 39
armpits, 1, 75, 177
Ascomycota, 59
Aspergillus fumigatus, 91
asthma, 130, 131, 175
athlete's foot, 66
atoms, 6
autoinducers, 114, 139

Bacillus fragilis, 204–205
Bacillus subtilis, 9, 11, 40–41, 163–164
Bacteria: vs. bacteria, use of terms, 1–2;
    cell characteristics from LUCA (last
    universal common ancestor), 25–30;
    communication (quorum sensing),
    13, 103–105, 138–139; detection and
    identification of (*see* detection and
    identification of microbes); DNA
    sequences database, 54–55; engulfed
    by eukaryote cells, 34–35, 141; four
    categories of shapes of, 40; genomes,
    9, 19–25; and horizontal gene trans-
    fers (HGTs), 20, 159–162; naming
    protocols, 5–6; numbers of, 30; as
    one of three major cell types, x, 2, 13,
    17–19; resistance to antimicrobials,
    157–163; structures of, 31(fig.); study-
    ing fossils of, 12–13; tree of life and,
    19, 20, 23–26, 27, 59–60; variability
    and adaptability of, 30–33, 157–163;
    viral infection of, 16, 124–125,
    159–160; viral infection of (bacterio-
    phages), 15, 16, 124–125, 159–160
"Bacteria first" hypothesis, 24
bacterial blooms: causes of, 178–180;
    in the digestive tract, 180–185
bacterial vaginosis, 126–127, 178,
    192–195
Bacteriodetes, 73, 75, 93, 124, 131,
    188–189
bacteriophages, 15, 124–125, 148,
    159–160

Basidiomycota, 59
B cells, 152–153
Beck, James, 129
Beijerinck, Martinus, 41
belly buttons (umbilicus), 82–85, 177
Berra, Yogi, 3
beta diversity, 50, 176–178
beta-globin protein, 168–170
Big Bang, 3
bilharzia (schistosomiasis), 166–167
biofilms, 112–116
bioluminescence, 138–139
birth: and belly buttons, 82–83; and
    puerperal fever, 134–136; types of
    delivery and microbiomes, 85–87
Black Death, viii
Blackwell, Meredith, 59
Blaser, Martin, 174
blpC gene, 9, 10, 11
B-lymphocytes, 151–153
bonobos, 108
Bosch, Thomas, 145
Brevibacterium spp., 79
broad-spectrum paradigm, 193
bubble mice experiments, 180–182,
    184–185, 186–191, 199–205
Buchnera spp., 21
bug, use of term, 1
Burkholderiaceae spp., 132

C (cytosine), 6–7
camel dung and dysentery, 163–164
cancer, 14, 175, 198–199
Candida spp., 104
Cano, Raúl, 16
capsids, 14–17
carbolic acid, 136
cell phones, 97
cells: characteristics, 25–30; human
    types of, 103; three major kinds of,
    x, 13, 17–19
chair microbiomes, 99–100
Chambers, John, 168

chemokine receptors, 141–142
childbed fever (puerperal fever), 134–136
childbirth. *See* birth
chimpanzees, 108
Chlorobacteria, 32
chloroplasts, 34–35
cholecystokinin, 196–197
chronic obstructive pulmonary disease (COPD), 130
circulatory system (blood): and immune system, 129, 148–149; lipopolysaccharides and inflammation, 184–185; malaria and sickle cell anemia, 167–170; sleeping sickness and kidney disease, 170–173; as sterile, 129
circumcision, 128–129
Clark, Arthur C., 18
classification. *See* naming and classification of things
classroom microbiome study, 101
cloning, 19–21
Clostridialies, 128
Clostridium spp., 199, 204–205
CODIS (Combined DNA Index System), 54
codons (code triplets), 8–11
coevolution and coexistence of microbes and hosts: benefits and risks of *Helicobacter pylori*, 173–176; and human ancestors, vii–viii, x; and hydrozoans, 146–147; malaria and sickle cell anemia, 167–170; sleeping sickness and kidney disease, 171–173
Cohn, Ferdinand, 40–41
colitis, 185, 203
comb of life, 20
communication, 4, 139–140, 142, 196–197. *See also* quorum sensing
conjugation mechanism of horizontal gene transfer, 160–162
convergence, 26–27, 34, 78–81

copepods, 66
Coprococcus, 79
coprolites (fossil feces), 121
core gene set, 21–22
core microbiomes, 106, 109, 131, 164, 177
corticosterone (ACTH, corticotropin), 199–200
corticotropin (ACTH, corticosterone), 199–200
Corynebacteria, 85
Corynebacterium, 74
cosmetics companies, 75
cows, fecal microbiomes of, 121–122
Crohn's disease, 203
Cryptomycota, 59
C-section births, 86–87
Cyanobacteria, 32, 40
cystic fibrosis, 130
cytosine (C), 6–7

Darwin, Charles, ix–x, 137, 157–158
databases, 54–55, 65, 104
DCA (deoxycholic acid), 199
Defense Enrollment Eligibility Reporting System (DEERS), 54
Deinococcus (spp.), 32
dental caries, 111–116
deoxycholic acid (DCA), 199
depression, 199–205
detection and identification of microbes, viii–ix, 39–63; culturing microbes, 41–45, 55–56, 59; discoveries made using culture-free approaches, 59–60; DNA databases, 54–55; DNA sequencing of human microbial communities, 58–60; early microscopes and, 1, 12, 13, 39–41; inference of existence and, 39; next-generation sequencing (NGS) approaches, 62–63, 67–68; PCR approach with clone sequencing, 57–58; rarefaction curve and, 60–62; using 16S ribosomal

detection and identification of microbes (*continued*)
RNA, 46–50; using DNA shotgun sequencing, 56–57, 61–62; using Sanger DNA sequencing, 50–55
Dicer protein, 147
dideoxynucleotides, 51, 53
Dietzia, 79
digestive tract, 117–125; altering gut ecology as defense against microbes, 163–165; anatomy of, 119; bacterial blooms and, 178–180; bubble mice experiments and, 180–182, 184–185; esophagus, 117; *Helicobacter pylori*, 173–176, 188; large intestine, 120; mucus layer functions, 182–184, 188–190, 203; rectum, 120–121; research difficulties, 117; small intestine, 118–119; stomach, 118. *See also* fecal microbiomes; gut-brain axis; obesity; oral cavity
diversity: alpha diversity, 50, 176–178; in belly buttons, 84; in digestive tract, 120, 121, 122, 123; in environments, 88, 89, 91, 94, 96, 97–98, 99–100; extent of microbial, ix, x; in the genital tract, 127, 194; of hand microbiomes, 69–71; in moist vs. dry skin regions, 73–75; sequencing and, 60, 61; of viruses and viromes, 64, 66
DNA (deoxyribonucleic acid), 6–11, 15–16
DNA barcoding approach, 48–50
DNA databases, 54–55
DNA sequencing: databases, 54–55; of human microbial communities, 58–60; and identification of criminals, 54; and identification of human remains, 53; next-generation sequencing (NGS) approaches, 62–63, 67–68; PCR approach with clone sequencing, 57–58; Sanger method, 50–53; shotgun analysis, 56–57, 61–62
dogs, 97–99
Drosophila melanogaster, 147
Dr. Strangelove, 69
DuBois, A. S., 136
Dunn, Rob, 84
dysbiosis, 178, 180, 194, 204–205
dysentery and camel dung, 163–164

ecological community analysis, 176–178
eczema, 77
Enterobacteriaceae, 5, 185
Enterococcus, 91
Enterotest, 117
environmental factors in microbe growth: changes in, and bacterial blooms, 179–180; communities' responses to changes in, 177–178; geography, 79–81, 99–101, 107–109, 123–124; humidity, 72, 74–75, 94, 129–130; and mouth microbiomes, 106; pH, 71, 105, 115–116, 118, 126, 177–178; temperature, 32, 72, 94, 129–130, 177–178; three major responses to changes in, 177–178; washing/cleaning, 70, 75, 95–96
environments, microbiomes of: classroom study, 101; houseomes, 94–99; Louvre Museum study, 91–92; offices, 81–82, 99–101; public restrooms, 92–94; subway studies, 88–91
Erb-Downward, John R., 131
erythrocytes (red blood cells), 149, 168–170
Escherichia coli, 2, 91, 185, 200
Escherichia spp., 118
Eukarya: cell characteristics from LUCA (last universal common ancestor), 25–30; interaction with

Bacteria, 34–35; and mutation, 159; as one of three major cell types, x, 2, 13, 17–19; structures, 34(fig.); unique characteristics of, 34

Evolution by Gene Duplication, 20–21

extremophiles, 13–14

Falkow, Stanley, 58, 67

fecal microbiomes: camel dung and dysentery, 163–165; cows and diet and, 121–122; fecal viromes, 124–125; in fossil feces (coprolites), 121; human, 121–125; and public restrooms, 92–94

fermentation, 191

fevers in children (study), 65–66

Firmicutes, 73, 75, 93, 95, 124, 131, 188–189

flagella, 33–34

Fleming, Alexander, 154–156

floors, 93

focal segmental glomerulosclerosis (FSGS), 171–173

fomites, 39

forced swim tests, 201–202

fossil bacteria, 12–13

fossil feces (coprolites), 121

fossil viruses, 16

Fracastoro, Girolamo, 39

FSGS (focal segmental glomerulosclerosis), 171–173

fungi, 59, 104

Fusobacterium spp., 132

G (guanine), 6–7

Gasteyer, Ana, 88

gastric cancer, 175

Gemella spp., 127–128

Genbank, 54

gender differences in human microbiomes, 70, 92–94, 99. *See also* genital tract

gene clusters (gene families), 22

gene duplication, 23

genetic code, 8–11

genetic variation. *See* variation and adaptation

genital tract: penis, 125–126, 127–129; vaginal microbiome, 86–87, 125–127, 178, 191–196

genomes, microbial. *See* microbial genomes

genome sequencing, 48

Genovese, Giulio, 171

genus, 5

geographic localities and microbiomes, 79–81, 99–101, 107–109, 123–124

GERD (heartburn), 175

germ, use of term, 1

ghrelin, 175, 196–197, 198

giardiasis (*Giardia lamblia*), 66

glucose, 197

Gogarten, Peter, 23

Gordon, Jeffrey, 123–125

Gould, Stephen Jay, 30, 64

guanine (G), 6–7

Guinea worm, 66

gut-brain axis: and appetite regulation, 196–198; brain development and the microbiome, 203–205; and depression, 199–202; gut membrane and bowel disorders, 203; and obesity, 198–199; and reactions to food, 196

gut microbiome. *See* digestive tract

Haemophilus spp., 132

hand microbiome study, 69–71

hand washing, 70, 134–136

Harvard University dental caries study, 111–112

Hawaiian bobtail squid, 138–139

health and definitions of healthy: bacterial blooms and sickness, 178–185; benefits and risks of *Helicobacter pylori,* 173–176; cancer and obesity,

health and definitions of healthy
(*continued*)
198–199; coevolution of malaria
and sickle cell anemia, 167–170;
coevolution of sleeping sickness and
kidney disease, 171–173; connecting
microbes to disease, 37–39; cultural
norms and, 166–167; depression and
the microbiome, 199–202; managing pathogens, 207–208; obesity,
diet and genetics, 186–191; vaginal
microbiomes and, 191–196
Helicobacter pylori, 118, 173–176
helper T cells, 151
hematopoietic tissue, 151
hemoglobin, 168–170
hemoglobin S alleles, 170
H-ESKD (hypertension-attributed end-
stage kidney disease), 171–173
Hewitt, Krissi, 99–101
HGTs (horizontal gene transfers), 20,
159–162, 178
Hilleman, Maurice, 137, 153
HIV/AIDS, 129, 158, 192
HMP (Human Microbiome Project),
ix, 38, 68, 71, 74, 76
HOMD (Human Oral Microbiome
Database), 104
Hooke, Robert, 39
horizontal gene transfers (HGTs), 20,
159–162, 178
houseomes (house microbiomes),
94–99
HPA (hypothalamic-pituitary-adrenal)
axis, 199–200
human body: cell types and internal
organs, 103; fears of microbe invasion of, 102–103; methods of microbe entry, 103–104; microbiomes
(*see* microbiomes)
Human Oral Microbiome Database
(HOMD), 104
humidity, 72, 74–75, 94, 129–130

hydrozoans, 144–146
hypersensitive response, 143
hypertension-attributed end-stage kidney disease (H-ESKD), 171–173

IBD (inflammatory bowel disease), 203
IBS (irritable bowel syndrome), 203
iChip culture method, 44–45
Illumina platform, 63
immune system: circumcision and,
128–129; genes and obesity and,
190–191; and lipopolysaccharides,
184–185; maternal immune activation (MIA) model, 204–205; Peyer's
patches and production of immunoglobulin A, 119; and skin microbes, 77. *See also* immune system,
acquired; immune system, innate
immune system, acquired, 150–157; cell
types and interactions of, 150–153;
learning and memory of, 153–154;
and overlap with innate system, 151;
vaccines and, 137–138; vertebrates
and, 150–151. *See also* immune
system, innate
immune system, innate, 140–150; a bacterium's experience with, 140–142;
complement system, 147–148;
complexity of, 138; differentiation
and the specialized cells of, 148–149;
evolution in lower animals, 143–150;
evolution in plants, 142–143; neutrophils and phagocytosis, 141–142;
recognition of self and, 138–140;
viruses and natural killer cells, 150.
See also immune system, acquired
immunoglobulins (Ig, antibodies), 119,
152–154, 184
indels (insertions-deletions), 24
inflammation, 203
inflammatory bowel disease (IBD), 203
insertions-deletions (indels), 24
interferons, 150

interleukins, 150
ion pumps, 30
irritable bowel syndrome (IBS), 203
Iwabe, Naoyuki, 23

"Jungle in There: Bacteria in Belly Buttons . . . " (Dunn, et al.), 84

keratinocytes, 72–73
keyboard studies, 81–82, 99
kidney diseases, 171–173
killer T cells, 151
kiss feeding, 107
kissing, 106–107
kitchens, 96–97
Klebisella spp., 118
Knight, Rob, 69
knockout mice, 185
Koch, Robert, ix, 41, 44
Koch's postulates, 44, 189, 200, 208
Kreitman, Marty, 173
Kyushu University study, 23

Lactobacillus rhamnosus, 200–202
Lactobacillus spp., 93, 111, 112, 118,
    126–127, 192–193
Langerhans cells, 128–129
large intestine, 120
last universal common ancestor
    (LUCA), 25–30
lectin receptors, 154
Lederberg, Joshua, 58
Leeuwenhoek, Antonie van, viii–ix,
    39–40
leptin, 175, 190–191, 198
leukocytes, 148–149
life: basic elements of, 6–11; core
    gene set, 21–22; defining, 17; gene-
    swapping and, 19–21; naming and
    classification of, 3–6; origins of, 3,
    11–14; shared characteristics of three
    forms of, 25–30; three forms of cellu-
    lar life, 17–19; viruses categorized as,

14–17; ways of tracing evolutionary
    history of, 22–25. *See also* tree of life
lipid chains, 28–30
lipopolysaccharides, 184–185
Lister, Joseph, 136
Louvre Museum study, 91–92
low-density microbes, 179–180
LUCA (last universal common ances-
    tor), 25–30
luciferase, 139
lungs, 129–133, 140–141
lymphocytes (T cells, B cells, natural
    killer cells), 149, 150, 151–153, 154, 191
lytic cycle/lytic phages, 124–125

Ma, Bing, 193–196
Madonna, 84
major histocompatibility complex class
    I (MHCI), 150
malaria, 37, 39, 167–170
Mason, Chris, 90–91
maternal immune activation (MIA)
    model, 204–205
membranes (cells), 28–30, 34
menstrual cycle and vaginal micro-
    biomes, 194–196
metazoans, 143–144
Methanosarcina spp., 21
MHCI (major histocompatibility com-
    plex class I), 150
MIA (maternal immune activation)
    model, 204–205
mice experiments, 180–182, 184–185,
    186–191, 199–205
microbe-associated molecular patterns
    (MAMPs), 145
microbes: adaptation and evolution
    of, 32–35; coevolution with hosts
    (*see* coevolution and coexistence of
    microbes and hosts); connecting
    microbes to disease, 37–39; defenses
    against, 70, 95–96, 134–136 (*see
    also* immune system); altering gut

microbes (*continued*)
  ecology, 163–165; detection and
    identification of (*see* detection and
    identification of microbes); distri-
    bution on human body, 67(fig.);
    diversity of (*see* diversity); focus
    on single pathogens vs. communi-
    ties of organisms, x; human body
    microbial communities, 38(fig.);
    human fear of invasion by, 102–103;
    parasitic organisms (*see* parasites);
    terms used for, 1–2. *See also* Ar-
    chaea; Bacteria
microbial genomes, 19–25; core gene
    set, 21–22; gene-swapping activities,
    19–21; tracing the evolutionary his-
    tory of, 22–25
microbiomes: complexity and vari-
    ability of, 66–68, 71; core, 106,
    109, 131, 164, 177; environment
    and (*see* environmental factors in
    microbe growth; environments,
    microbiomes of); gender differ-
    ences, 69–70; Human Microbiome
    Project (HMP), ix, 38, 68, 71, 74, 76;
    viromes, 64–66. *See also* digestive
    tract; genital tract; lungs; skin
    microbiomes
microscopes, 1, 12, 13, 39–40
Mimivirus spp., 14
Mineo, Sal, 83
mitochondria, 34–35
molecular clock method, 16–17
mouth. *See* oral cavity
MUC2 gene, 185
mucus layer functions (digestive tract),
    182–184, 188–190, 203
Mullis, Kary, 51
multipotent hematopoietic stem cell,
    148–149
mutations, 28, 158–159, 168–169, 171,
    181–182, 190
mutualistic relationships, 161

myeloid cells, 149–150
myeloid precursor cells, 151

naming and classification of things,
    3–6
National Center for Biotechnology
    Information (NCBI), 54
National Institutes of Health (NIH),
    67–68, 71
National Library for Medicine, 54
natural killer cells, 150, 154
natural selection, 157–159
Nature, 58
NCBI (National Center for Biotechnol-
    ogy Information), 54
Neisseria spp., 112
neomycin, 157
neutrophils, 141–142, 143, 147, 148,
    149–150, 191
next-generation sequencing (NGS)
    approaches, 62–63, 67–68
NIH (National Institutes of Health),
    67–68, 71
NOD-like receptors, 145
North Carolina State University studies,
    84, 94–99
nuclear membranes, 34
nucleotides, 6–7

obesity: Bacteriodetes to Firmicutes
    ratio, 124, 188–189; and cancer, 198–
    199; and diet and genetics, 186–191;
    H. pylori, leptin, and ghrelin and,
    175; lipopolysaccharides and, 185
office environments, 81–82, 99–101
Ohno, Susumo, 20–21, 22
oligofructose, 189–190
Olsen, Gary, 46
On the Origin of Species (Darwin), x,
    137, 157–158
oral cavity, 104–110; core microbiome
    of, 106; geographic and phylogenetic
    components of, 107–109; kissing

and, 106–107; microbe species and variability, 104–105; as open system, 105–106; periodonitis, 116–117; teeth (*see* teeth); tonsils, 109–110

orexigenic reactions, 197, 198

organ transplants, 131, 132

Pace, Norman, 56–57, 88–89

Pandoravirus spp., 14

panspermia hypothesis, 11–12

paralog rooting, 23

parasites: bacteria genera *Buchnera* and *Wolbachia*, 21; bacteriophages, 15, 16, 124–125, 159–160; bilharzia (schistosomiasis), 166–167; malaria, 167–170; numbers of, 66; plasmids, 161; trypanosomes, 170–173

Pasteur, Louis, ix, 39, 41, 136

Pasteurellaceae, 110

pathogenicity islands (PGIs), 162–163

PathoMap Project, 90–91

pattern-recognition receptors, 142–143, 145

PCR (polymerase chain reaction), 14, 51, 57–58

penicillin, 154–156

peptic ulcer disease, 175

periodonitis, 116–117

Petri dishes, 42–44

pets, 97–99

Peyer's patches, 119

PGIs (pathogenicity islands), 162–163

pH, 71, 105, 115–116, 118, 126, 177–178

phages. *See* bacteriophages

phagocytosis, 141–142

phones, 97, 99–100

photophores, 139

phyla (phylum), 73

phylotypes, 73–75, 84

pillowcases, 97

pilus, 160–161

plant immune systems, 142–143, 154

plasmids, 161–162

Plasmodium falciparum (malaria), 37, 39, 167–168

polymerase chain reaction (PCR), 14, 51, 57–58

polyploids, 22

Porphyromonas gingivalis, 117

Porphyromonas spp., 132

Prevotella intermedia, 112

Prevotella spp., 116, 126, 127–128, 132

Princess Bride, The, 206–207

probiotics, 189, 200–201, 203, 204–205

Prokaryotes, 18, 26

Propionibacterium spp., 74

proteins, 7–11

Proteobacteria, 73, 75, 93, 95, 116, 131

Pseudomonas spp., 132

public restroom microbiomes, 92–94

puerperal fever, 134–136

quorum sensing, 3, 113–115, 138–139

rarefaction curve, 60–62

RDP (Ribosomal Database Project), 55

RdRP (RNA-dependent RNA polymerase) gene, 65

rectum, 120–121

red blood cells (erythrocytes), 149, 168–170

Redi, Francesco, 39

Red Queen hypothesis, 125

Relman, David, 58, 67

Rendezvous with Rama (Clark), 18

resilience, 177–178

resistance, 76–77, 157–163, 177

restroom microbiomes, 92–94

Ribosomal Database Project (RDP), 55

ribosomal RNA (rRNA), 46–50, 55

ribosomes, 28, 46

Richards, Thomas, 59

RNA (ribonucleic acid): characteristics, 6–11; "RNA world" hypothesis, 12; 16S ribosomal RNA, 46–50, 55, 56–58; and viruses, 15–16

RNA-dependent RNA polymerase (RdRP) gene, 65
RNAi pathway, 147
Roche 454 platform, 62
roller derby study, 77–81
Rothia dentocariosa, 113
rRNA (ribosomal RNA), 46–50, 55, 56–58

Sagan, Carl, 3
Salmonella enterica serotype typhimurium, 161
Sanger, Frederick, 50–53
Sargasso Sea microbial communities analysis, 58
schistosomiasis (bilharzia), 166–167
Schopf, J. William, 13
Selenomonas spp., 112
Semmelweis, Ignaz, 134–136
September 11, 2001 attacks (identification of remains), 53
sexually transmitted infections, 192
shoes, 97
sickle cell anemia, 168–170
Sinkkonen, Aki, 83
16S ribosomal RNA, 46–50, 55, 56–58
skin microbiomes (Skin Microbiome Project), 71–82; babies' acquisition of, 85–87; belly button (umbilicus) study, 82–87; dry regions, 75; environmental changes and, 75–76; four basic phyla of, 73; geographic localities and, 79–81; hand microbiome study, 69–71; and identification of individuals (keyboard study), 81; moist regions, 74–75; pathogenic threats to, 76–77; roller derby study, 77–81; sebaceous regions, 74; shedding of microbes, 89, 101 (*see also* environments)
sleeping sickness, 170–173

small intestine, 118–119
smallpox, 137
smokers, 131–132
Sneathia spp., 127–128
species, 5
spontaneous generation, 39
Spurlock, Morgan, 186–187
squames, 72–73
Staphylococci, 85
Staphylococcus aureus, 77, 161–162
Staphylococcus epidermidis, 76–77, 113
Staphylococcus spp., 74, 110, 154
star stuff (stardust), 3
stomach, 118, 196–205
stop codons, 9–10
Streptococcus mutans, 111–112, 113, 115
Streptococcus spp., 77, 91, 110, 113, 116, 132
streptomycin, 157
subway studies, 88–91
Sudo, Nobuyuki, 199–200

T (thymine), 6–7
TAd island (tight adherence island), 162–163
Tanner, Anne, 111–112
Tannerella forsythia, 112, 117
T cells, 151, 153, 191
teeth, 110–117; dental caries, 111–116; structures of, 110–111
temperate phages, 124–125
temperature, 32, 72, 94, 129–130, 177–178
Texas A&M University, 97–99
Thermus aquaticus, 13–14
thymine (T), 6–7
tight adherence island (TAd island), 162–163
Time, 136
toilets, 93, 95, 96, 97
toll-like receptors, 145

tonsils/tonsillitis, 109–110

transduction mechanism of horizontal gene transfer, 159–160

transformation mechanism of horizontal gene transfer, 159

transpeptidase, 156

Travis, J., 183

tree of life, x, 18–19, 20–21, 23–25, 26–27, 30

Treponema denticola, 117

trypanosomes, 170–173

Turnbaugh, Peter J., 188

twins, microbiomes of, 106, 123–124

type II diabetes, 175

Tyson, Neil deGrasse, 3

U (uracil), 6–7

umbilicus (belly button), 82–85, 177

University of Arizona office study, 99–101

University of California, San Diego, office study, 99–101

University of Colorado studies, 69, 92–94

University of Exeter pond water study, 59

University of Idaho vaginal microbiome studies, 193–196

University of Maryland vaginal microbiome studies, 193–196

University of Michigan, Ann Arbor, smoker's microbiomes study, 131–132

University of Oregon studies, 77–81, 101

uracil (U), 6–7

urine microbiomes, 127–128

U.S. military DNA database, 54–55

vaccines, 137–138

vaginal microbiome, 86–87, 125–127, 178, 191–196

vagus nerve, 197, 202

Valko, Emery I., 136

variation and adaptation: conjugation mechanism of, 20, 160–162; humans and, 31–32; mutation and natural selection and, 157–159; transduction mechanism of, 20, 159–160; transformation mechanism of, 20, 159; and the tree of life, 32–35

Veillonella spp., 128, 132

Venter, Craig, 58

Veterans Affairs Health System, Michigan, 131–132

Vibrio fischeri, 139

viral paleontologists, 16

viromes, 64–66, 124–125

viruses: bacteriophages, 15, 124–125, 159–160; genomes, 64–66; and mutation, 158; origins of, 14–17; and transduction, 159–160

Waksman, Selman, 156–157

washing/cleaning, 70, 95–96, 134–136

Washington University, St. Louis, research, 123–124, 188

Weill Cornell Medical College, 90

Wilson, E. O., 2

Winogradsky, Sergei, 41–42

Woese, Carl, 13, 46

Wolbachia spp., 21

work places. *See* office environments

Xanthomonas spp., 79

zombies, 102